EXPLORATION OF PALMER ARCHIPELAGO

ANTARCTIC BIRDS
Ecological and Behavioral Approaches

David Freeland Parmelee
Foreword by Harold F. Mayfield

University of Minnesota Press
Minneapolis   Oxford

Published by the University of Minnesota Press
2037 University Avenue Southeast, Minneapolis, MN 55414

Printed in the United States of America on acid-free paper

Library of Congress Cataloging-in-Publication Data

Parmelee, David Freeland, 1924-
    Antarctic birds : ecological and behavioral approaches / David
Freeland Parmelee.
      p.   cm.
    Includes bibliographical references.
    ISBN 0-8166-2000-8 (hc)
    1. Birds—Antarctic regions—Ecology. 2. Birds—Antarctic regions—
Behavior. I. Title.
QL695.2.P368 1991
598.2'5'26210989—dc 20   91-12378
                                             CIP

A CIP catalog record for this book is available from the British Library.

I dedicate this book to the late Professor George Miksch Sutton, who introduced me to the Arctic; to Dr. George A. Llano, who introduced me to Antarctica; and especially to Jean and Helen Gale for their courage, sacrifice, and devotion on the home front.

CONTENTS

FOREWORD

The Antarctic is a region of mystery. Indeed, it is the end of the earth. Yet, it is a region of exceptional interest, not only because of its remoteness, but also because it brings together the incomparable richness of the polar waters and the incomparable desolation of the antarctic land. Few of us will ever visit it, and so we need an observer with a sharp eye to tell us about it. Here we have the best possible guide, David Parmelee, an experienced scientist who is also an accomplished artist and photographer. For this purpose he does not focus on the desolate mountains and glaciers of the interior as some visitors do, but gives us a view of the Antarctic through the window of the Palmer Peninsula and adjacent islands, where the ocean waters meet the land.

His qualifications for this work are exceptional. Blessed with a strong physique, he is well fitted for physical exertion under difficult conditions, and he has brought to this task a deep understanding of life in the related but strikingly dissimilar Arctic.

I first met David Parmelee more than twenty years ago in the Arctic, where his name was already a legend. Unlike many scientists who visit remote places in the spirit of explorers, traveling through and never coming back, David made each study site his own, returning again and again in following years until he was an authority on it. It is impossible to write a scholarly report on the birds of the Arctic without quoting his researches.

When his professional responsibilities at the University of Minnesota required his residence in that state during the season of the short arctic summer, he turned to the other end of the world at the other end of the year. Again, he applied his talents in his own special way, selecting promising sites and concentrating until he was the authority on them. Like a true professor, he has opened this part of the world to his students and, indeed, to all of us.

Harold F. Mayfield

PREFACE

Long before that moment in 1972 when I first stepped onto Antarctica, the seeds of my love for polar regions had been sown and had flowered. Professor George Miksch Sutton, my graduate school mentor and a seasoned polar ornithologist, planted these seeds through his writings and paintings of northern birds and Inuits. When he hired me as his field assistant and whisked me off to Frobisher Bay—that great icy indentation of the south coast of Baffin Island just beyond Canada's Hudson Bay—these seeds were cultivated, nurtured, and took deep roots. Here was a world that fulfilled one's wildest dreams: a land reverberating with strange bird calls throughout an endless summer light, a land touched with lichens and flowers and miniature forests teeming with lemmings and owls, a land scribed by char-filled streams, all racing toward a sea rimmed by towering cliffs with colorful raptorial aeries. I found its marine environment no less attractive: a frigid sea that spawned countless invertebrate forms, fascinating marine mammals and myriad eiders, a sea that boasted of flood tides so high that the daily schedules of native and visitor alike were governed by its movements and moods.

Nothing in all the world that I had experienced up until then had come close to Frobisher Bay, and I embraced Sutton's polar world as my own. Then, for nearly two decades, I thought only Arctic. I spent my time either joining or organizing expeditions to the far north, with Sutton participating in several of them. If there were new birds and habitats to explore, always I looked northward beyond timberline, never south. My perspective ended abruptly when I was hired by the University of Minnesota to direct its field biology program involving two field stations. This would be a professional advancement that could not be dismissed lightly. My summers would be spent at the university's Forestry and Biological Station at Lake Itasca, a part of Minnesota not too different from Michigan's Upper Peninsula, where Jean, my lifelong companion and spouse, and I grew up. Nevertheless, the decision to join the University of Minnesota was difficult and traumatic. It has often been said that there is no rose without thorns: my rose was a challenging summer period at Itasca; the thorns were the severe curtailment of my arctic work.

Once I was established in Minnesota, it became apparent that while my summers were taken up with administrative duties, my winters were open for research. Only then did I think Antarctic. My life had become topsy-turvy, so why not consider the other end of the world, with its pleasant austral summer that also basked in continuous polar light—a fact that had long been engraved in the chromosomal structures of our bipolar migrants, the Arctic Terns (*Sterna paradisaea*).

Although it was Sutton who introduced me to the Arctic, it was his former Cornell colleague Dr. George A. Llano who introduced me to the Antarctic. A new world was about to open up: the incredible continent of Antarctica, with its vast ice shelves and encompassing pack-ice ecosystem with polar birds all foreign to me. Llano was the chief scientist for the Division of Polar Programs of the National Science Foundation. Dr. Donald B. Siniff, a University of Minnesota colleague and an authority on antarctic seals, introduced me to Llano. Within months of that meeting, I was flying off to Antarctica, with hardly enough time to prepare for my first encounter with the world's incredible seventh continent. The more I read about it, the more I was convinced that no place on earth was its equal, not even my beloved Arctic.

ACKNOWLEDGMENTS

The field studies were financed by the following Division of Polar Programs (National Science Foundation) Grants-in-Aid:
GV 36032, DPP 7421374, DPP 7615350, DPP 7722096, DPP 8213688, and DPP 8715630. Many units and individuals assisted my students, colleagues, and me directly or indirectly in various ways too numerous to mention. Those individuals whose influence and expertise had a profound impact on the success of our research were G. A. Llano, F. S. L. Williamson, A. F. Betzel, and C. E. Myers of the Division of Polar Programs; P. J. Lenie, master of RV *Hero*; and G. M. Jonkel, chief, U.S. Bird Banding Laboratory. H. F. Mayfield of Toledo, Ohio, reviewed and edited the manuscript critically and made numerous helpful suggestions. To the many units and individuals who provided administrative, logistic, and scientific advice and assistance, I heartily extend sincere thanks, in chronological sequence as follows:

National Science Foundation, Division of Polar Programs:
Division Heads J. O. Fletcher, A. N. Fowler, R. H. Rutford, E. P. Todd, P. E. Wilkness, and their staffs, with special thanks to W. R. Seelig, G. G. Guthridge, P. Lewis, Jr., K. N. Moulton, D. M. Bresnahan, R. B. Elder, B. T. Fogle, R. L. Cameron, B. Lettau, D. M. Anderson, R. B. Williams, H. Holloway, Jr., A. Inderbitzen, T. E. DeLaca, P. A. Penhale, and S. Draggon. The author and the University of Minnesota Press gratefully acknowledge financial support from the Division of Polar Programs for the publication of this book.

National Science Foundation, Public Affairs Office:
Special thanks to R. Kazarian.

University of Minnesota, College of Biological Sciences:
Deans R. S. Caldecott, D. C. Pratt, H. B. Tordoff, C. T. Magee, and their faculties and staffs, with special thanks to D. B. Siniff, J. Jarosz, R. Oehlenschlager, C. Ray, R. M. Schaefer, R. Kaercher, D. F. McKinney, D. Warner, P. Snustad, G. R. Murdock, D. T. Luce, K. T. Williams, E. Hanson, D. Joyce, A. M. Fosdick, D. M. Berube, T. G. English, S. L. Frye, A. L. Will, D. K. Bromenshenkel, G. B. Ownbey, K. Larntz, J. Tester, R. Phillips, J. T. Locke, K. Winker, G. E. Nordquist, G. A. Voelker, J. T. Klicka, R. D. Benson, and especially D. E. Gilbertson, former director of the Bell Museum, and E. Birney, current director. Very special thanks to S. J. Neumeister, who typed innumerable drafts of the manuscript, and to F. J. Cuthbert for reviewing the manuscript.

University of Minnesota, Office of Research and Technology:
Assistant Vice President A. R. Potami and staff, with special thanks to J. M. Garland, J. A. Krzyek, M. Bergass, and R. Dunn.

Antarctic Support Services, Holmes & Narver, Inc.:
Project Managers R. J. Buettner and R. L. Murphy and their staffs, with special thanks to E. L. Herbst, M. C. Lanyon, K. L. Worthing, R. J. Wolak, E. Myioda, and L. Myioda.

National Museum of Natural History, Washington, D.C.:
Special thanks to G. E. Watson, J. P. Angle, and R. C. Banks.

USCGC *Glacier*:
Captains W. E. West, C. P. Gillett, and W. P. Hewel, their officers and crew, with special thanks to J. McClellund, K.-P. Hsu, J. Coste, B. Genez, P. L. Hagstrom, and D. Meekins. Participating scientists were T. Foster, T. Kvinge, A. Foldvik, R. Bo, F. Todd, and B. Obst.

Argentina:
Special thanks to A. Blomquist, J. Bomfiglio, M. A. E. Rumboll, J. Gallardo, and J. C. Godoy of Buenos Aires, and T. and N. Goodall of Ushuaia.

RV *Hero*:
Captains P. J. Lenie and N. Deniston, their officers and crew, with special thanks to J. Polkinghorn, M. Malcahy, W. Church, D. Martin, K. Hauk, M. Levine, and G. Dobrosky. Participating scientists were R. Hoffman, N. Flesness, D. Siniff, R. Reichle, F. Gress, B. W. de Lappe, L. R. Schock, I. Cameron, B. Obst, K. Nagy, and V. Komarkova.

U.S. Palmer Station:
Project Managers/Deputy Directors E. R. Koenig, R. J. Wolak, A. Brown, M. Eichenberger, and their staff, with special thanks to L. Jukkola, G. E. Bennett, W. M. Lokey, M. Sturm, C. Patrick, S. Williams, W. F. Lincoln, E. P. Gilmore, W. Tofani, E. Myioda, L. Myioda, R. Zogbaum, D. Gaffeney, J. Fields, D. B. Wiggin, D. R. Mortvedt, S. R. Dame, M. Snyder, G. M. and R. J. Heimark, and P. Jorgensen. Participating scientists were D. Murrish, B. Davis, N. K. Temnikow, R. Moe, D. Condit, E. Porter, B. Daniels, D. L. Laine, W. J. Showers, R. R. Di Paola, J. Edwards, P. Moriarty, P. C. Tirrell, C. Denys, J. Chisholm, J. Katsufrakis, W. Curtsinger, B. Merdsoy, J. Baust, R. Lee, J. Warburton, R. R. Veit, K. Nagy, B. Obst, A. Clarke, W. Coats, J. F. Heinbokel, and J. Jolman.

British Antarctic Survey:
Director R. M. Laws and his staff, with special thanks to W. Sloman, W. N. Bonner, R. Chinn, B. Block, E. Harvey, R. Herbert, D. Fletcher, J. Priddle, G. Picken, R. Herbert, J. Conroy, M. Pawley, R. Hanks, A. Smith, D. McKay, J. M. Gallsworthy, I. Robarts, R. Birnie, P. A. Prince, B. Pearson, and J. P. Croxall.

RRS *John Biscoe*:
      Captain E. M. S. Phelps and his officers and crew, with special thanks to P. Clarke-Halifay, D. Bray, M. Shakesby, R. Wade, H. Binnie, G. Cutland, and A. Baker.

Falkland Islands:
      Special thanks to D. R. Morrison, I. Jones, D. J. Sollis, I. J. Strange, C. D. Kerr, J. H. McAdam, and S. and J. Poncet.

HMS *Endurance*:
      Captains N. Bearne and D. L. Deakin and their officers and crews, with special thanks to C. Errington, P. Hurst, N. Crocker, A. Swain, and S. Hill.

MS *Lindblad Explorer*:
      Captain H. Nilsson and his officers, crew, and staff, with special thanks to F. Erize, K. Shackleton, O. S. Pettingill, Jr., R. T. Peterson, and D. Bartlett.

ITT Antarctic Services, Inc.:
      Directors R. E. Gray and R. A. Becker and their staffs, with special thanks to A. Brown, R. Broglie, M. C. Lanyon, and D. B. Wiggin.

University of Minnesota Veterinary Diagnostic Laboratories:
      Special thanks to D. M. Barnes.

University of Minnesota Photographic Laboratories:
      Manager T. Pangborn and staff.

University of Minnesota Space Science Graphics:
      Special thanks to M. W. Bossen, C. Faust, B. Szurek, and C. Swanson.

University of Minnesota Press:
      Former Director J. Ervin and former Assistant Director W. Wood, with special thanks to B. J. Kaemmer, R. N. Taylor, and C. Masters. Very special thanks to current Director L. Freeman, natural science editor B. Coffin, production manager K. Wolter, and P. Gonzales, who handled the publication of this book. Thanks also to D. Mathers, the designer, and to J. Selhorst, my copy editor.

Zoologisk Museum, Copenhagen:
      Special thanks to the late F. Salomonsen and J. Fjeldsa.

Percy Fitzpatrick Institute of African Ornithology/Marion Island:
      Director W. R. Siegfried and his staff, with special thanks to J. Cooper, R. Abrams, R. K. Brooke, S. Fugler, R. Cassidy, B. Watkins, T. Salinger, P. Condy, P. van Litsenborgh, R. Riley, and V. Smith.

SA *Agulas*:
      Captain W. Leith and his officers and crew, with special thanks to H. Bergh.

Chile:
      Special thanks to R. P. Schlatter of Valdivia, and P. C. Avila and H. Valencia of Santiago.

Point Reyes Bird Observatory:
      Director D. G. Ainley and staff, with special thanks to W. Z. Trivelpiece.

Travel Dynamics, New York City:
      Special thanks to Expedition Leader D. Schoeling; also to A. Sexton, C. Gifford, L. Myers, D. Malelu, P. Bradley, J. Renda, B. Hart, M. Munn, S. Weierbach, P. Guttman, J. Hammond, S. Lenz, D. Carballo, H. Dawson, J. Marshall, and R. Polatty.

MV *Illiria*:
      Captains A. Parisis and I. Pittas, their officers, especially C. Ramadas, and crew; also Ice Master P. J. Lenie. Participating naturalists were D. Cameron, E. Carriazo, S. El-Sayed, R. W. Matthews, C. A. Morgan, H. Zomer, W. L. Sladen, M. Sallaberry, R. Schlatter, G. Webers, J. Valencia, C. Marangunic, R. Polatty, J. Rabassa, A. Schiavini, and A. Mazzotta. Participants from the National Audubon Society were S. Drennan and M. Drennan. Participants from the American Museum of Natural History were S. W. Hardee, S. Quinn, S. Clift, and K. Chambers.

Polar Circle:
      Captain I. Slettevoll and his officers and crew. Participating naturalist was K. Winker.

Minnesota Science Museum:
      D. Chittenden and staff.

Also, special thanks to the late D. Nethersole-Thompson and M. Nethersole-Thompson of Ross-shire, Scotland, and to A. Petersen of the Iceland Museum of Natural History in Reykjavik.

INTRODUCTION TO ANTARCTICA

The Incredible Seventh Continent

Antarctica connotes the coldest, windiest, driest, certainly the most monochromatic iced-over place on earth. It possesses the most of nearly everything that one finds inhospitable. Coming out of Antarctica are incredible temperature recordings that exceed -88°C. Catabatic winds cascading down its many glacial tongues at velocities exceeding 200 kilometers per hour are terrifying. The coupling of these low temperatures with high winds produces windchills beyond belief. We are told that Antarctica is our greatest desert, where the mean annual precipitation is often less than 12 centimeters, all of which falls as snow. Not even snow falls in the Dry Valleys of Victoria Land, where no plants grow in the soil surfaces. There is no dearth of ice, however; 98 percent of Antarctica's 14.3-million-square-kilometer land surface is covered with it up to 5,000 meters thick in places—enough ice to cover the United States with a layer 3,704 meters high, or roughly 2 miles thick.

So much ice covers Antarctica that its enormous weight depresses the land beneath it. Some estimates suggest that half the continent lies below sea level, despite the fact that all of this ice also makes this continent the world's highest overall. Coastlines around the world would be altered dramatically should all this ice melt.

If these superlatives are not enough to influence your impression of Antarctica, consider the most talked about current concern. Straddling the seventh continent is a big "ozone hole" that threatens to let environmentally damaging ultraviolet radiation escape from its usually trapped state in the ozonosphere.

Among the valid conceptions popularizing Antarctica today are many misconceptions. Foremost among them is a widespread belief that the continent is uniform throughout, purely rock and ice, all gray and white, surrounded by ice-choked seas that remain frozen most of the time and inhabited by black-and-white penguins. If this were true, all of the continent's birds and other wildlife would probably be uniform throughout as well, but this we know is not the case. Even the penguins are widely but unevenly distributed in the Southern Hemisphere, though many of us think of them as being strictly synonymous with Antarctica. Most penguin species and their numbers or biomass occur on oceanic islands outside of Antarctica, in a more hospitable zone called the Sub-Antarctic. Many also reside on the other three southern continents whose waters abut the great Southern Ocean.

Few bird species, including penguins, are able to adapt to the rigorous physical constraints of Antarctica. Most of those that do cling to the periphery of the continent proper, where continuous ice cliffs extend indefinitely, and where relatively few exposed headlands and offshore islands provide snow-free breeding sites from one season to the next. The most ameliorable coastal areas occur along the Antarctic Peninsula, which projects 1,600 kilometers northward from the main continent in a 45-260-kilometer-wide curve between 63° and 75° S latitude. Differences are pronounced between its eastern and western coasts, even in this limited area of the continent. The peninsula's considerably colder eastern coast is permanently fringed by pack ice and an ice shelf that is not nearly so suitable to wildlife as is its west coast north of 68° at Marguerite Bay.

Not only is the peninsula's northwest coastal region one of the continent's most prolific areas for life forms, it is perhaps its most scenic from the standpoint of pristine, rugged terraine. A mountain range originating in the Andes extends an arm far out into the South Atlantic Ocean before turning back and eventually forming the very backbone of the Antarctic Peninsula. The

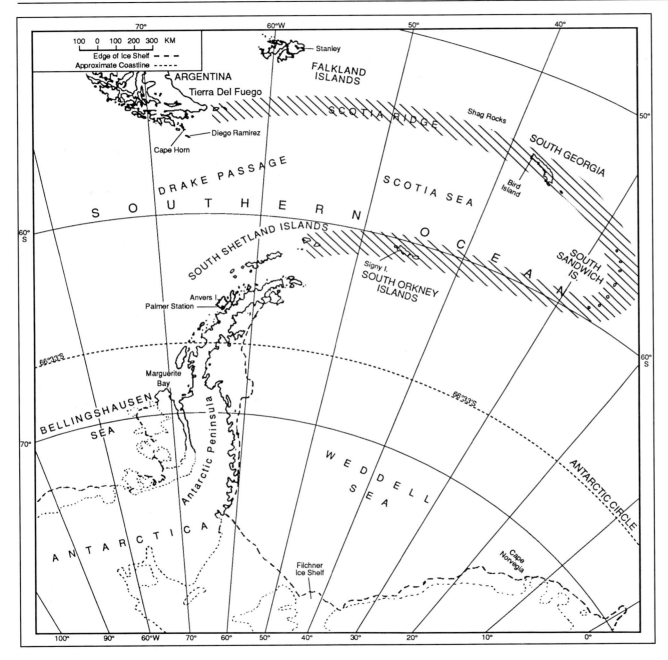

Figure 1. Map showing the relative positions of South America, Scotia Ridge, and the Antarctic Peninsula. Note the location of Palmer Station on Anvers Island off the northwest coast of the Peninsula.

mostly hidden submarine range referred to is the Scotia Ridge (see Figure 1). It has only a few emerging peaks, but these form several notable island chains or archipelagos. Those lying fairly close to the Antarctic Peninsula are the South Sandwich, South Orkney, and South Shetland islands. One that closely parallels it geographically, physically, and biologically is Palmer Archipelago, which lies at the center and heart of this study (Figure 2). Here the Andes resume their lofty heights, rising straight out of the sea to elevations of 2,800 meters or more, forming a succession of ice-cliffed fjords and spontaneous active glaciers. Contrasting boldly with the shimmering whites and deep intrinsic blues of the glaciers are the formidably black volcanic intrusions of ancient Cretaceous and Jurassic times. Other factors contribute to the region's unusual beauty.

The Antarctic Peninsula region, including much of the South Sandwich and all of the South Orkney and South Shetland islands, is classified as "Maritime Antarctica," based largely on its plant life related to annual precipitation. Dessication may well be the limiting ingredient for plants in this greatest of all deserts. As one may surmise, plants, in terms of species and bio-mass, decrease with increasing latitudes, a situation comparable to that of birds. While algae dominate the marine and fresh waters, mosses and lichens command the land, though, surprisingly enough, one can also find two species of flowering plants, but only in the maritime zone. Wherever one finds birds concentrated, one also finds a copious growth of plants. The two constitute a marriage. No bird colony would be complete without its colorful ground and cliff sites decorated with botanical blacks, greens, yellows, and oranges. A monochromatic scene? Hardly.

Precipitation rules the day in many areas of the world, and no less so in Antarctica. As expected, precipitation in the maritime zone decreases with increasing latitudes, starting from 400 to as much as 800 millimeters annually in the north to 200 millimeters or less in the south. These differences are thought to be due in part to the strong westerly winds in the north, with their mean annual velocities approaching 30 kilometers per hour, while those in the south are less than 20 kilometers per hour under a milder influence called the East Wind Drift. Mean temperatures also decrease with increasing latitudes, from

-13.6°C in the north to -16°C in the south during the coldest month of the year. Temperatures above freezing occur frequently in summer, the January mean being pronounced.

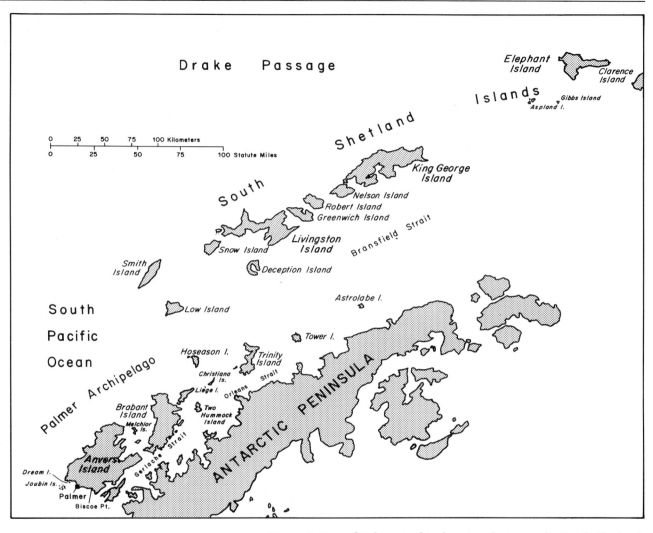

Figure 2. Map of Palmer Archipelago in relation to the South Shetland Islands and the Antarctic Peninsula. Tower Island lies at its northern extremity, while Anvers Island and the Joubin Islands close by are at the southern end. Note the narrow straits that separate Palmer Archipelago from the Antarctic Peninsula.

In the following chapters much will be said about the sea ice that influences birds and the people who study them. Sea ice constitutes a major physical and biological system, sometimes referred to as the pack-ice ecosystem. The formation of pack ice is a short-lived annual event—unlike the perpetual freshwater ice sheets that cover the continent, and then push their leading edges far out to sea until broken free as gigantic flat-topped icebergs. During the polar winter, fresh pack ice derived from salt water builds layers 2-3 meters thick, only to break up the following spring. Much of it disappears during the summer months, except in unusually cold areas, where it may persist for longer periods, as is often the case on the Weddell Sea or on the eastern side of the Antarctic Peninsula. Each year so much pack ice forms, breaks up, and forms again all around Antarctica that it is sometimes referred to as the "pulsating continent." At waxing maximum, Antarctica's pack-ice collar covers an incredible 20 million square kilometers (an area larger than the continent itself); it then shrinks at waning to a mere 3 million square kilometers (Fogg and Smith 1990).

Although the winter pack extends upward of 280 kilometers northwest of the Antarctic Peninsula, for reasons not fully understood the area west of the peninsula, including Palmer Archipelago, does not freeze completely shut every winter. Broad areas of the pack open and close periodically, not necessarily in the same places within and between seasons. Some years are more open than others. Ornithologically, this region is highly unusual, if not unique, for Antarctica, for certain birds can and do reside there year-round, while the non-breeding visitors find an exceptional winter feeding ground. Not the land but the sea is the fundamental source of food for all birds residing in Antarctica. Even the terrestrial bird predators are only one step removed, for the penguins and other species they prey on are totally dependent on the sea.

First-time visitors to Maritime Antarctica will do well to consider the advantages and disadvantages of both northern and southern sectors. Generally, in the South Orkneys and South Shetlands, one will experience somewhat warmer temperatures than those farther south. Access to these northern areas is comparatively easy because of ice-free sailing conditions, which are at times not present farther south. Exposed land areas are more extensive during summer in the north, presenting not only more explorable areas, but also easier logistics. On the other hand, the higher winds of the north bring with them more clouds and precipitation. Cloudless, sunny days increase southward, enhancing the majestic, breathtaking scenery of the southern sector.

Palmer Archipelago lies somewhere between the extremes of Maritime Antarctica's northern and southern sectors. I took long sailing voyages to get there, but once there I knew I had found the study site of my favorite subject—polar birds.

ANTARCTIC BIRDS

The Discovery of a Polar Study Area

U.S. Palmer Station on Gamage Point, Anvers Island. South Polar Skua on Bonaparte Point overlooking Hero Inlet in foreground. Photographed 26 November 1975.

George Llano assigned me to an extended cruise on an icebreaker aboard the U.S. Coast Guard Cutter *Glacier*, with S. D. Mac-Donald of Canada's National Museum of Natural Sciences in Ottawa as my field companion. We were asked to critique George E. Watson's newly written manuscript dealing principally with field identification of birds of the Antarctic and Sub-Antarctic. MacDonald and I had spent six months in 1955 on Ellesmere Island in the Canadian High Arctic. I knew that he was a superb field biologist capable of picking up minute details of flying birds even at great distances. On a rolling ship in rough seas the conduct of at-sea observations is serious business; I knew of no better person to assist me in putting George Watson's field guide to a trial test at sea.

Stu MacDonald and I flew by military aircraft via New Zealand to U.S. McMurdo Station in the Ross Sea area of Antarctica. At McMurdo we recorded our first antarctic birds—South Polar Skuas (*Catharacta maccormicki*), sometimes referred to as McCormick's Skuas, and the bird species that occurs farthest south. Before leaving McMurdo shortly after Christmas, we flew to Cape Royds for our first look at nesting penguins reputed to be the most southern of their kind. They proved to be Adélie Penguins (*Pygoscelis adeliae*) attending chicks at nests built of stone, all within the shadow of a smoking Mount Erebus, Antarctica's most famous volcano. Other than the skuas that partitioned the penguin colonies for eggs and chicks, we saw only a few Wilson's Storm-Petrels (*Oceanites oceanicus*)—one of the very widely distributed and conceivably the most abundant of seabirds that summer in Antarctica. As is the case of northern birds in the Arctic, southern birds decline in numbers of species toward the higher latitudes. McMurdo was too sparse a breeding ground and nursery for the bird studies I had in mind.

Once the USCGC *Glacier* broke free of the compacted pack ice of McMurdo Sound, she cruised comfortably in the nearly wide open waters of the Ross, Amundsen, and Bellingshausen seas. For the next week or so, we slipped past gigantic flat-topped icebergs and crunched through occasional strands of a fast-receding pack, recording the much sought-after seabirds as we went. After crossing the South Pacific's sixtieth parallel on a northward swing away from the southern seas, we had our first indoctrination to the Drake Passage—that churning, spindrift stretch separating Antarctica from South America and aptly dubbed the wildest sailing waters around. Few aboard the *Glacier* disagreed. One can credit the Drake for being a great place for the big albatrosses that excel in wind and grace the lonely seas. Our destination was far eastward, deep within Antarctica's Weddell Sea, but we diverted from a direct course there to take on fuel at a Chilean port in the Strait of Magellan. During the refueling operation, the *Glacier*'s twin helicopters on several occasions

dropped MacDonald and me off at Isle Contramaestre and Isle Magdalena, which are renowned for their Magellanic Penguins (*Spheniscus magellanicus*), Chilean Skuas (*Catharacta chilensis*), and a host of other seabirds (Parmelee and MacDonald 1975). Quite unexpected, however, were nesting ibises within meters of occupied penguin burrows. Here was a strange mix of species in a fascinating part of the world, certainly worthy of a student project, but for me the higher latitudes took precedence.

The Weddell Sea is a vast ocean basin that feeds extremely cold waters into the South Atlantic. It also is an icy place on the surface, so much so that even our modern ships are not always able to penetrate its pack, in western parts especially. The 1972-73 season was, I suppose, highly favorable for such travel, for the *Glacier* went further south in the Weddell than any surface ship previously. Punching through extraordinarily dense pack on occasion, it finally moored in fast ice beyond the seventy-seventh parallel within view of the impenetrable Filchner Ice Shelf. While Norwegian and U.S. scientists busied themselves with projects involving deep ocean currents, *Glacier*'s twin helicopters ferried MacDonald and me along the shelf for hundreds of kilometers. The only birds seen except in open water were small groups of Emperor Pen-

guins (*Aptenodytes forsteri*), a handful of South Polar Skuas, and a single Southern Giant Petrel (*Macronectes giganteus*).

The complete void of exposed land areas was disconcerting to an ornithologist bent on finding a prolific study area. All along this vast ice-shelf edge of the Weddell Sea the only local breeders were the Emperor Penguins, all perfectly adapted to a nesting ground of snow and ice throughout the austral winter, with its inhumane windchills in near total darkness. True, a few species such as the Antarctic Petrels (*Thalassoica antarctica*) and Snow Petrels (*Pagodroma nivea*) seek out the isolated nunataks or mountain peaks that protrude through glacial ice; nevertheless, these birds must fly back and forth to the open sea to feed themselves and their chicks, in some cases for many kilometers. In all our Weddell travels we spotted but one tiny nunatak; its smooth contours lacked ledges and the craggy cracks so requisite for nesting petrels. At trip's end MacDonald and I were convinced that although the Weddell Sea has a restricted breeding ground, its pack-ice ecosystem is an indispensable feeding ground for many seabirds. It certainly is a significant molting and staging area for migrating Arctic Terns (Parmelee and MacDonald 1973; Parmelee 1977a;

Zink 1981b). At the time, the birds were readying themselves for the long flight northward into the path of those fierce winds coming out of the Drake. In tabulating our bird counts for the Weddell Sea, we found that Arctic Terns were the third most abundant species encountered. Since then I have not recorded a single one outside the Weddell Sea during subsequent expeditions to Antarctica.

I was pleased to tell George Llano that both MacDonald and I gave high marks to George Watson's field guide, and later I published a favorable review (Parmelee 1977b). I also related to Llano our experiences overall. Especially the Weddell Sea, with its incredible physical and biotic interwebbings, was everything a polar biologist could hope for. Still, it lacked a certain ingredient essential to my plans, namely, a readily accessible study area suited to the research interests and capabilities of our University of Minnesota graduate students. Llano quickly responded by putting me and one of my ablest students, Stephen J. Maxson, aboard the National Science Foundation Research Vessel *Hero*. Its captain was Pieter J. Lenie, a skillful master of that little vessel, which was destined to be an integral part of our program.

In November 1973, RV *Hero* sailed from the port of Ushuaia, Tierra del Fuego, Argentina, and headed east along the Beagle Channel, with albatrosses and giant petrels trailing in its wake. With Cape Horn at our starboard, we headed into the infamous Drake in all its stormy fury and, if my field notes or the lack of them are any indication, my traveling companions and I experienced an unwitting lost weekend. No one stood up, much less appeared for meals, which were prepared by a chef who was undaunted by the mighty Drake.

Robert Hoffman and Nate Flesness researched seals; they wished to explore the mostly uncharted waters of Stigant Point, a rugged coastal area on the north side of King George Island in the South Shetland Islands, a location that even our seaworthy Captain Lenie viewed with misgivings. We found fur seals—not many—but enough to score as a foothold here for a species making a comeback following an era of uncontrolled slaughter. Higher on the slopes were large numbers of Chinstrap Penguins (*Pygoscelis antarctica*), some already trying to incubate eggs in persistent snows that still covered their traditional stony nests.

Following several shore landings at Stigant Point, we pushed farther west along the island chain to Nelson Island, where, at Harmony Cove, the anchorage was fairly good and the shore landing easy. We saw large numbers of several species of birds breeding there, and I knew that my quest for an antarctic study area was coming into focus.

RV *Hero* gained latitudinal heights by following a southwest course from the South Shetlands into the more frigid zone of Palmer Archipelago—a long chain of ice-capped islands that parallels the great Antarctic Peninsula. Soon to become a familiar sight to my students and me was Gerlache Strait, a watercourse that separates the two regions. Its usual open, mostly ice-free sailing lanes were used by many a whaler and explorer, and the route remains popular today. From there, when the sea is not shrouded in water clouds, the spectacular mountains of the Antarctic Peninsula come into view. In giving us a close-up surveillance of the peninsula, Captain Lenie charted the ship's course along a stretch of its western flank known as the Danco Coast, with shore landings at Waterboat Point and Paradise Bay. The latter accomodated a nicely appointed Argentine station, Almirante Brown, perched on jutting black rocks above a pounding surf and, it seemed to me at the time, precariously close to massive glaciers. Ours was the first ship in that season and we brought with us a huge backlog of mail, new faces, and new talk for a very patient group of staff members and scientists who had just wintered over under severe isolation. In return for these amenities, the Argentines put on a wine and beef feast the like of which we had not experienced before or since. It was easy to figure why Almirante Brown was a favorite stopping-off point for ships enroute to Anvers Island, the site of U.S. Palmer Station, some 40 nautical miles away near the Palmer Archipelago's southwestern extremity.

My search for a study area concluded upon our arrival at Palmer Station. Practically at the station's doorstep was a little offshore island overrun with thousands of Adélie Penguins whose excrement turned its exposed rocks and remaining snows krill red. Seals snoozed complacently on floes within a stone's throw of the station's little wharf, where South Polar Skuas and Kelp Gulls (*Larus dominicanus*) wheeled overhead. Ribbons of flying Antarctic Blue-eyed Shags (*Phalacrocorax atriceps*) returning from feeding forays moved rhythmically toward some great rookery. What really excited me was a milling cloud of Antarctic Terns (*Sterna vittata*) that rose

suddenly nearby when several skuas flew over. Here was a little-known polar species that was readily accessible from a U.S. station equipped with field equipment, laboratories, and living accommodations.

Steve Maxson and I soon discovered that we could not walk from the station to the tern colony. A fairly wide inlet that emerged from beneath Anvers Island's ice cap separated the point of land on which the station was located from the one occupied by the terns. We had to use a boat to get from the station's Gamage Point across Hero Inlet to Bonaparte Point, where the colony was situated, or use a hand trolley on a long steel cable that had been stretched across the inlet by the U.S. Navy Seabees who had built the station. Despite some of its shortcomings, we usually opted for the trolley system. Once ashore on Bonaparte Point, I arrived at an important decision: National Science Foundation permitting, for the next several years Palmer Station would be the heart of my study area. That same day I received approval to leave RV *Hero* and put up at Palmer Station, and with that done Steve Maxson and I initiated our study of Antarctic Terns—the first of many projects to follow.

In the beginning my goal simply was to study the ecology and behavior of a few select species of charadriiform birds, not-

ably skuas, gulls, and terns, since that group related closely to my arctic work. Habitats immediately adjacent to Palmer Station were especially attractive, not only because of accessibility to a rich avifauna, but also because many of the birds resided there year-round, one of few such places in the Antarctic. Moreover, both the South Polar Skua and Brown Skua (*Catharacta lonnbergi*) bred sympatrically in a narrow zone of overlap that included Anvers Island, affording an unusual opportunity to observe the two under similar environmental conditions. Virtually nothing was known about the other charadriiform species at high southern latitudes. Although our studies centered on the charadriiformes, some of my students developed a keen interest in other species as well. Southern Giant Petrels and Antarctic Blue-eyed Shags proved to be especially good subjects to study, and our goals broadened by including them.

Vital to our studies from the start was the banding (ringing) and color coding of breeding birds. In searching for banded individuals, including those that disperse widely upon reaching breeding maturity, our travels took us far beyond the confines of our original Palmer study area. In time we accumulated a mass of information on

Emperor Penguins deep within the Weddell Sea. Photographed 2 February 1973.

all species of birds along Palmer Archipelago's many waterways and island coasts (interiors covered with ice except for high mountain peaks). Not all areas of the archipelago were visited in this study, notably the extreme northern sector, where some of the region's earliest bird observations were recorded (e.g., Andersson 1905). In recent years, Poncet and Poncet (1987) have recorded a number of seabird colonies at Tower Island as well as in northern Trinity Island and a few other places not seen by us.

We were encouraged to conduct these broad surveys by the participating nations involved internationally with the Scientific Committee on Antarctic Research (SCAR), of which I was a U.S. representative on one of its subcommittees. The logical conclusion to my expanded long-term goal is this monograph on the birds of Palmer Archipelago.

Obstacles and Compromises

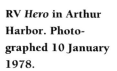

RV *Hero* in Arthur Harbor. Photographed 10 January 1978.

olar ice in all its intricate forms is the fundamental component of the antarctic environment. Polar ice is the drawing power that intoxicates and traps explorers, adventurers, tourists, and scientists who seek unspoiled outdoor laboratories. Polar ice is also the greatest impediment to their varied pursuits, and more often than not dictates harsh terms regarding their transportation and supply. Palmer Station, as I was about to discover, is not an easy place to get to or from by conventional transport. The smooth glacier sheet that rises ever so gently behind the station's physical plant gives the illusion of an ideal landing surface for Twin Otters or other small fixed-winged aircraft, but is in fact cut insidiously throughout by crevasses, many hidden by mere crusts of snow. Planes landing there do so at considerable risk, and this is a practice not condoned by the National Science Foundation. The only feasible access to

Palmer is by sea. Even then, surface vessels have to confront unpredictable pack ice that, when sufficiently compacted, forestalls even the most carefully planned arrivals—a facet of polar life demonstrated by our repeated attempts to reach the Palmer study area earlier than usual.

Annual relief of personnel overwintering at Palmer Station happens with the arrival of the first ship at the start of the austral summer season. This usually takes place sometime between late November and mid-December, when pack-ice conditions are judged to be reasonably good. Late arrivals prolong the agony of the personnel overwintering and anticipating an early homeward departure, but they also hamper the work of those researchers contemplating fieldwork based on phenology of the subjects they study. Ornithologists need to be on the breeding ground early. By the time Steve Maxson and I arrived at Palmer in late November 1973, the Antarctic Terns and other seabirds were far along in their breeding cycle. With this in mind, I made special arrangements for an early arrival at Palmer the following season. The British Antarctic Survey would deliver me there on one of its ships headed for Palmer Archipelago.

RRS *John Biscoe* cruised unhampered down the Gerlache, entered the gorgeous Neumayer Channel cut through mountains and ice, discharged men and supplies onto an island called Doumer, then proceeded to Palmer through what soon became known to my students and me as the dreaded Bismarck Strait. Only 10 nautical miles out from Palmer, the ship bogged down hopelessly in thick pack ice, and with its predicament vanished all expectations for an early arrival. For me there was but one recourse: return with *John Biscoe* to the Falkland Islands and recoup. And recoup I did, but not until I had spent considerable time at South Georgia with *John Biscoe* and a British ice reconnaissance ship, the HMS *Endurance*. Back at the Falklands for the third time that year, I finally boarded the tourist ship MS *Lindblad Explorer* and, fol-

lowing multiple shore landings at South Georgia and other places, finally arrived at Palmer Station on the eve of Christmas— fully two months after having first joined the *John Biscoe* in Uruguay.

The next season started no better. In our second October attempt to reach Palmer early, it was RV *Hero* this time that got stuck in the Bismarck, but, fortunately for Bill Fraser and me, only a few nautical miles out from the station. Palmer's winter staff came zigzagging through fields of rubble ice and escorted us in on skis. Among the smiling welcomers was Dave Neilson, the first of my students to overwinter and who was determined to see the entire breeding cycle of his chosen subjects, the South Polar and Brown skuas. The year-long stay ensured his being at the Palmer study area when the birds first arrived and last departed. Overwintering, it seemed to us, was the solution to Bismarck's pack—it was a bad assumption.

Other than the hand trolley that ferried us almost anytime to Bonaparte Point, travel throughout the remainder of our study area was contingent on outboard engine-powered rubber boats known to us as "zodiacs." These proved to be marvelous craft, capable of high speeds on open water to be sure, but also very good at close-in

maneuverings and pushing through loose pack and achieving difficult landings on rocky shores in slapping surfs. But when southerly offshore gales slammed Bismarck's pack into Palmer's Arthur Harbor for days on end, zodiacs became impotent.

I cannot soon forget those horrendous times when the harbor was beset with ice. Dave Neilson and I struggled to guide our zodiac along narrow cracks that momentarily spread open when we shoved aside a loose floe with an oar. Failing in that, we hopped atop the unsteady floes and dragged the rubber craft over the really bad stretches. A zodiac stint in open water to the islands of Litchfield and Humble was accomplished within minutes; in compacted ice, minutes were translated into hours. So exhausting and time-consuming was this futile procedure that Dave Neilson finally packed up his bird bands and camped among the skuas of Litchfield, a practice adopted later by Neil Bernstein, who resided among the shags of Cormorant Island.

If we had tough going with Bismarck's pack, consider the plight of the birds we studied. Surface-feeding species were highly dependent on open water for their prey foods. Prolonged, unfavorable ice conditions resulted in failed or even aborted breeding efforts from the start (Parmelee et al. 1978). Fish-dependent South Polar Skuas were particularly sensitive to foraging opportunities, and this was clearly reflected in the number of young produced; peak years in fledgling production were followed by those in which productivity plummeted to zero. Only recently has it become apparent that the extent of winter sea ice is highly variable in the Palmer area, with severe ice conditions occurring roughly every seven to ten years and lasting two to three years. If this holds true, boom and bust years are predictable for certain species. Other species fared better. Penguins, for example, overcame these conditions by diving through cracks in the pack and pursuing krill and other foods far beneath the surface. Penguin productivity may be influenced by changes in food stocks, but it appeared to be unaffected by ice conditions during the breeding season, as seemed to be true also of the Brown Skuas, which depended on penguin eggs and chicks for the survival of their progeny. Polar ice had little jurisdiction over the breeding biology of these species linked advantageously in the ecological food chain.

We now know that the undersurface of pack ice is a haven for myriad simple life forms that contribute much to a food chain that spirals upward to the skuas and other predators at its apex. This same pack that maintains an abundance of life can be hard on other kinds of organisms, however. It keeps them in check through a surf-propelled, abrasive, ice-scouring action along all the rocky shores that it leans upon and polishes clean. Such conditions were visible on some of the exposed islands that punctuated the sea off Palmer, where naked shores supported few invertebrates. One notable exception was the tenacious Antarctic Limpet (*Nacella concinna*), which at all seasons somehow managed to overcome both surf and pack. In turn, the Kelp Gulls had adapted perfectly to a year-round fare of limpets, except for a brief period in chick rearing when fish becomes an important supplement. Our second overwintering researcher, Bill Fraser, was fascinated by this gull-limpet association. Among Fraser's many important observations was the discovery that gull distribution in Antarctica parallels that of the limpet.

Polar ice helps keep the seas frigid throughout the austral summer, and it is this cold water that one fears when crossing the pack on foot or working out of zodiacs. Once submerged, a human has but a few minutes to live, as we are told repeatedly. The stringent rules on safety laid down by RV *Hero*'s Captain Pieter Lenie and the National Science Foundation's "Buzz" Betzel can be credited to the relatively few close calls at sea experienced by my group. At Palmer one takes note of ice and wind forecasts. Nevertheless, field biologists are often uptight with phenology. Squeezed in the dilemma between safety and work output, sometimes they push aside all rules and common sense. So it was the day Dave Neilson and I attempted to wind up observations on Hermit Island before the long zodiac ride back to Palmer. We failed to heed soon enough a sharp change in the weather and, well out from shore, suddenly became helplessly enveloped by a dense fog, in wind and waves far exceeding zodiac standards. The gale swept us several kilometers off course, but luckily back against the Anvers Island glacier rather than out to sea, where our remains would surely be today. Incidents of this sort prompted the establishment of caches of food and survival gear on many of the islands off Palmer Station. We thought it far better to land at last-chance shores than to be cast at sea.

Fickle winds and a pack on the move usually signaled change. An empty Arthur Harbor one moment was suddenly filled with ice the next, and atop many of the floes came seals of several species. One floe-loving creature that fascinated my Palmer colleagues was the Leopard Seal (*Hydrurga leptonyx*). Its sinister serpentine head with formidable jaws was a topic of great concern. Marine divers especially gave this seal a wide berth. On occasion I stretched out comfortably in a zodiac and kept a watchful eye for any approaching Leopard Seal to allow the divers down below to do their studies. Whenever I spotted one of the seals I simply turned on the outboard engine to alert the divers, who responded by popping to the surface and concluding their diving for the day. Whether these fierce-looking seals had a propensity for attacking humans remained a moot question. Some felt as I that their threat was grossly exaggerated; others were convinced that it was simply a matter of time before we lost one of Palmer's diving biologists.

During our innumerable encounters with Leopard Seals, we had but one disturbing incident. A fairly large individual several times bumped the bottom side of our zodiac and momentarily frightened Bill Fraser, Dave Neilson, and me. Naturally,

we headed straight for Breaker Island nearby and scrambled up its precipitous front with the boat's landing rope in hand. The seal hardly looked our way, but while we watched from above, it viciously attacked the zodiac and mouthed the craft's rubbery surface. For whatever reason, fortunately, it did not clamp down on the tubing. Still, a single thought came to the three of us: the real threat lay in being tossed from a tooth-punctured zodiac into a sea of polar ice. Since then, several zodiacs (none occupied at the time) have been punctured by Leopard Seals at Palmer Station (P. Jorgensen, personal communication).

Polar ice, as we have seen, is the principal cause of logistical and safety problems for the biologist at sea. It poses less of a problem on land unless one has to cross crevassed ice sheets, something my students and I have not chanced. The rocky peninsulas and offshore islands within our Palmer study area were, however, partially covered with snow and ice, and precautions taken while walking and climbing made sense. Slippery surfaces resulted in bruises and sprains, but in nearly all cases they were no more serious than those inflicted

daily by Palmer's numerous dive-bombing skuas in defense of nests. Pamela Pietz miraculously survived a painful fall down one of Litchfield Island's higher cliffs. Following two major operations for injured vertebrae, she later returned to Antarctica on a midwinter cruise, displaying unusual fortitude and dedication to polar work.

Quite intentionally I have used these examples not to portray romantic adventure or human endurance, but rather to make the point that research in remote polar regions is time-consuming and requires many seasons. Researchers working under controlled laboratory conditions in readily accessible sites cannot possibly appreciate the problems associated with polar ice. Our greatest contribution to polar biology was that our observations spanned a fair number of seasons and included a multitude of banded birds with known histories. Even so, we all agree that our combined efforts merely scratched the surface for research opportunities in Palmer Archipelago and elsewhere along the Antarctic Peninsula.

Leopard Seals on an ice floe in Arthur Harbor near Palmer Station. Photographed 22 January 1975

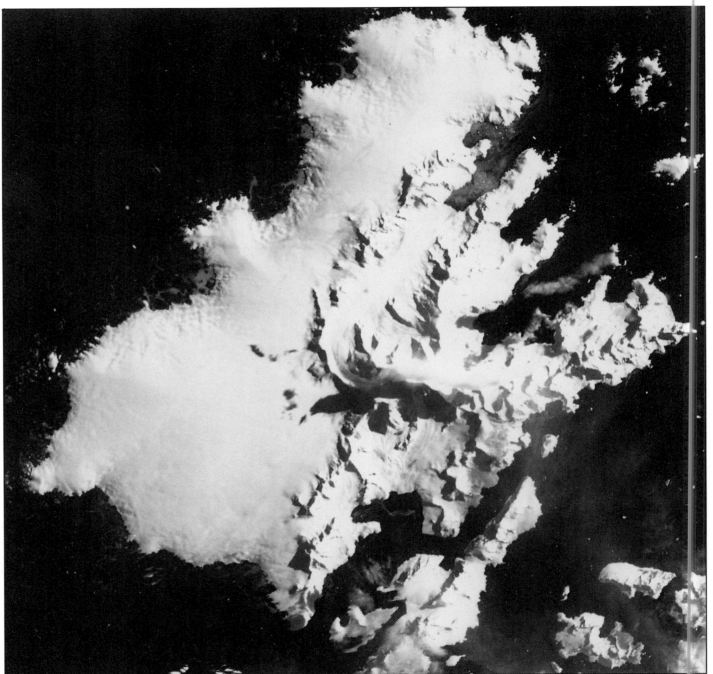

Conservation Imperative

C

Satellite picture showing the high mountains and extensive ice cover of Anvers Island. U.S. Palmer Station is located about midway on its south coast. Also shown near the south coast are the smaller Doumer and Wiencke islands.

Considering the vastness of Antarctica's polar ice, it is easy to see why so few inhabitable areas exist for its wildlife. Paradoxically, the great Southern Ocean that produces an abundance of food provides a paucity of breeding space for its land inhabitants. Even the relatively snow-free Sub-Antarctic has a very limited breeding ground, for most of its islands are small and thinly scattered over a huge body of water. The crowded real estate occurs not only on the beaches where seals haul out and penguins converge, but also along the traditional trails that lead to breeding sites. A few more options are open to birds endowed with flight, but often these amount to the precipitous fronts of sea cliffs or far inland nunataks. Because of these highly crowded conditions, the territories of these birds are very small and rigidly defended nesting spaces. An incubating bird's territory in Antarctica often does not extend much beyond a stretched neck and beak.

No small advantage afforded these far-southern birds and mammals is their nearly complete freedom from land predators over the millennia. Skuas and other traditional predators exact a relatively small toll of eggs and chicks on the breeding grounds, compared with those that fledge. Surviving young are thought to encounter their greatest losses during their first seaward venture or migration, though little is known about how and where this comes about. Whatever the biological factors involved, the result has been the evolution of highly uninhibited land-breeding vertebrates. The word that best describes this nearly unparalleled behavior is *innocence*—primitive innocence.

The first explorers/exploiters to reach Sub-Antarctica, and later Antarctica, found a colossal windfall. There were seals beyond belief, and all stacked conveniently along the beaches and primed for the inevitable slaughter. They found also penguins so innocent that they waddled into the jaws of straining, tethered dogs. Penguins, too, were herded to the great rendering pots that were designed in the beginning for seal blubber, but later were made larger for the whales. Ubiquitous camp followers, rats and cats, have been particularly hard on burrow-nesting petrels and other bird species in the Sub-Antarctic. Despite the terrible inroads made on these bird and mammal populations, to this day the survivors still possess the primitive innocence of their ancestors. Not enough time has elapsed for these vertebrates to adapt to a fast-changing environment controlled by humans. It is difficult to anticipate this kind of innocence without firsthand experience; for instance, compared with antarctic seals, the seals in the Arctic are not at all naive and consequently are extremely difficult to approach—doubtlessly because of their long association with bears and later with native peoples. My photographic record of seals is a fair indication of these regional differences: I have a handful of arctic seal photos, poor ones at that, compared with countless good ones of antarctic seals.

Sea captains look for negotiable harbors in which to spend polar nights. These are the very places where Antarctica's wildlife assembles for breeding purposes, and where scientists usually build their research stations. Casual visitors love to see both the wildlife and the stations, and are not often turned away. It is conceivable that varied human activities will exploit Antarctica and squeeze its concentrated wildlife into remnants of its potential in the absence of stringent conservation policies. Unless excessive ultraviolet radiation spoils the Southern Ocean's basic food stocks, or rising global temperatures start an irreversible melting of the great ice sheet, it is not too late to save Antarctica from wholesale deterioration. However fragile Antarctica's wildlife, it has the resilience to bounce back. A reassuring sight at many abandoned stations these days is the reoccupancy of nesting and loafing sites by former inhabitants. Penguins and other species are prone to nest all around deserted buildings, and they will even occupy the interiors of these structures should a door blow open. Unfortunately, the problem goes deeper than simple penguin reclamation.

Too many of these abandoned stations are little more than junk piles, truly eyesores that detract from Antarctica's pristine beauty. Because of the good anchorages and other safety features, tour ships often visit these abandoned sites. Conservation-minded passengers aboard ship become highly indignant toward the nations that leave their unsightly messes behind. Some visitors, however, delight in exploring historic sites, whatever their condition. An opinion voiced frequently is that the abandoned stations judged to have historic value should be preserved, and the rest should be removed. Compromises will have to be made on this particular subject. There are other concerns to be compromised as well.

Growing numbers of visitors left unchecked pose a serious problem not only for Antarctica's fragile environment, but also for the scientific community. As a research scientist, I had to leave my field studies to go aboard visiting ships and lecture to groups. Now I enjoy lecturing to groups while continuing my research on an informal basis aboard cruise ships. My students were always highly vocal about their particular concerns, especially when the birds they were studying would suddenly leave their territories and fly off to some approaching ship, where they were fed by the passengers and crew. Artificial feeding of birds introduces a serious bias in scientific studies; more important, it disrupts the ecology and behavior of the birds per se. It gives predators, for example, an undesirable advantage over prey species. An effort is currently being made to close the open dumps at stations and to discourage the feeding of birds, including the indiscriminate dumping of garbage at sea.

Worse than the feeding of birds is the disruptive influence people have on wildlife when approaching nesting individuals too closely—a great temptation in Antarctica, where it is easy to interpret innocence for friendliness, which it is not. A frightened elephant seal becomes virtually a bulldozer when running amok among nesting penguins. Gulls are prone to fly far from their eggs and chicks in the presence of humans, allowing the less inhibited skuas to fly in for an easy kill. Particular attention must be afforded the plants, for the delicate mosses and lichens recover very slowly from trampling underfoot. Examples of this sort are too numerous to list, but it is clear that inexperienced people need guidance when visiting an unfamiliar breeding ground. Modern cruise lines have trained personnel conduct their groups ashore, for their leaders now realize that it is all too easy to affect adversely many living organisms, even during one brief visit on a limited breeding ground.

Crowded conditions have an adverse effect on the scientists working in Antarctica as well. Unless projects are monitored and supervised carefully within and between seasons, scientists can unknowingly interfere with or damage other sci-

entists' research. Especially susceptible are long-term bird-banding and population studies. Scientists are not beyond reproach. Along with everyone else, they must heed conservation policies apart from their professional pursuits.

Ships have been grounded in polar ice until sunk, but until recently there had not been a major oil spill in Antarctica. On 28 January 1989 our worst fear was realized when the Argentine *Bahia Paraiso* entered Arthur Harbor and struck an underwater ledge that ripped a 10-meter gash in its side. At the time, the ship was transporting about 450 200-liter drums of diesel fuel and 217 canisters of compressed butane and propane gas. Fortunately for the passengers and crew, Palmer Station was only 3 kilometers away, and not one person was lost. Unfortunately, on Torgersen Island nearby were thousands of young Adélie Penguins poised for their first swim at sea.

Initially, a 1-square-kilometer fuel slick formed near the disabled *Bahia Paraiso*, but within a few hours it had spread to the beaches of Torgersen Island and beyond. On 31 January 1989, the ship broke free from its original grounding, drifted, and finally sank near DeLaca Island, spilling more fuel, which eventually covered the entire Palmer study area and beyond. At the time of the accident William R. Fraser was

stationed at Palmer Station, while I was aboard the cruise ship MV *Illiria*, which participated in the rescue operation. Following the initial impact of the accident, one thought foremost in our minds was the possible loss of our long-term study, which involved thousands of man-hours. Count our single study among the many carried on at Palmer and the total loss is staggering, not to mention the environmental damage to one of the choicest biological sites in all of Antarctica.

Bird losses known or presumed to have resulted from the Arthur Harbor spill are mentioned in the main text of this study dealing with individual species. Although it will be some time before meaningful conclusions regarding the spill will be available from those working currently at Palmer Station, bird assessments should be quite conclusive largely because of the area's many banded birds with known histories.

To preserve certain antarctic areas from excessive human usage, a fair number of sites have been identified for their unique physical and biological attributes and designated as Specially Protected Areas (SPAs) under the Antarctic Treaty, which is recognized by many nations round the world. One SPA within the Palmer Archipelago is Litchfield Island, which also lies within the

Palmer study area. Other areas not quite so restrictive are designated as Sites of Special Scientific Interest (SSSIs). Currently there is a move to provide area-wide protection for Palmer Station—probably a spin-off from the 1989 spill. One possible Marine SSSI in Palmer Archipelago under consideration is at Dallmann Bay, off the west coast of Brabant Island. Whatever the designation, official permission is required to conduct research, let alone to set foot on, any of these protected areas. The Permit Office, Division of Polar Programs, National Science Foundation, Washington, D.C., accommodates Americans who qualify. Others must obtain permits from their respective nations.

I have mentioned a few threats related to Antarctica's fragile environment, but certainly there are others. Increasing visits by many interests other than tourism are cogent concerns, notably research and supply ships, aircraft, private seagoing recreation vessels, commercial exploitation of krill (a basic antarctic food stock), potential fishery, and petroleum and mineral industries. These are subjects currently being discussed by various international committees. The Antarctic Treaty holds our greatest promise for a universally sound management policy for the entire antarctic region south of the sixtieth parallel. A few countries lay claim to parts of this immense region, but many feel, as I do, that the Antarctic Treaty will prevail for a long time, if not indefinitely.

Biologists view birds as a major component of the environment and particularly good subjects to study from the standpoint of several scientific disciplines, including ecology, behavior, and physiology. Because they are highly sensitive to adverse changes in the environment, birds are extremely good indicators of even small changes that otherwise easily escape detection. Purely from the standpoint of biology much could be said to justify the investment of both time and money in research devoted to them. Knowledge of birds purely from the standpoint of human pleasure more than justifies research efforts as well, for few vertebrate groups have commanded such a wide interest generally. All this notwithstanding, my colleagues and I also take the view that the most important justification for such research is simply the conservation of a valuable resource—based on the premise that the more we know about a species, the better the chance for its survival within a framework of enlightened policies and management.

Landing sites in Antarctica often are at a premium: Adélie Penguins at one of the oft-used launching pads at Hope Bay, Antarctic Peninsula. Photographed 20 December 1974.

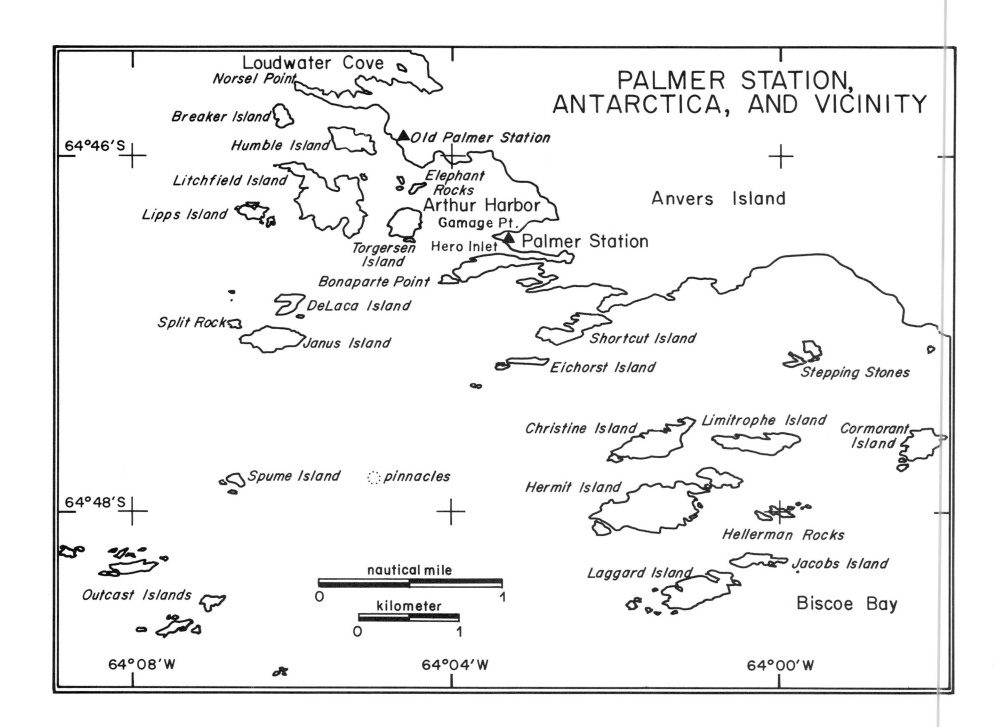

PALMER STATION,
ANTARCTICA, AND VICINITY

Loudwater Cove
Norsel Point
Breaker Island
Humble Island
64°46'S
▲ Old Palmer Station
Litchfield Island
Elephant
Rocks
Arthur Harbor
Anvers Island
Lipps Island
Gamage Pt.
Hero Inlet ▲ Palmer Station
Torgersen
Island
Bonaparte Point
DeLaca Island
Split Rock
Janus Island
Shortcut Island
Eichorst Island
Stepping Stones
Christine Island
Limitrophe Island
Cormorant
Island
Spume Island   :::pinnacles
Hermit Island
64°48'S
Hellerman Rocks
Laggard Island
Jacobs Island
nautical mile
Outcast Islands
0   1
Biscoe Bay
kilometer
0   1

64°08'W
64°04'W
64°00'W

NATURAL HISTORY OF THE PALMER STUDY AREA

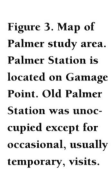

Figure 3. Map of Palmer study area. Palmer Station is located on Gamage Point. Old Palmer Station was unoccupied except for occasional, usually temporary, visits.

he Palmer study area included two prominent peninsulas of Anvers Island, Bonaparte and Norsel points, and groups of islands extending from Cormorant Island in the east to Breaker Island in the west (see Figure 3). The area was defined arbitrarily by a 5-kilometer radius extending seaward from Palmer Station, for that was the prescribed safety margin for zodiacs. Travel beyond the study area required the use of large craft. As previously noted, there was a remarkable assemblage of birds that bred or visited this particular part of Palmer Archipelago (Parmelee et al. 1975). The area is also highly attractive to other disciplines in the biological and physical sciences, as amply demonstrated by the 357 publications that have resulted from Palmer Station-related research and later compiled for a report produced at a workshop held at the University of California, Santa Barbara (Quetin and Ross 1988).

Anvers Island was named in 1898 by the Belgians (under Adrien de Gerlache) for the Province of Anvers, Belgium. Arthur Harbor was first charted from 1903 to 1905 by the French (under Jean B. Charcot), who named Bonaparte Point for Prince Roland Bonaparte, then president of the Paris Geographical Society. The British were the first at Arthur Harbor to establish a research station, which they built in 1955 on Norsel Point, named after their ship *Norsel*. Among the British faunal surveys conducted there emerged a report on the area's birds for the years 1955-57, including important winter records (Holdgate 1963). During this period the British also named a number of geographic sites within the Palmer area. Arthur Harbor was named for Oswald R. Arthur, then governor of the Falkland Islands. Litchfield Island was named for Douglas B. Litchfield, then general assistant and mountaineer at the station. Torgersen Island was named for Torstein Torgersen, then first mate on *Norsel*. Janus Island at the entrance to Arthur Harbor was named for the ancient Roman deity who was guardian of the gates. Hermit Island was so named because a Britisher "spent some time on this island alone" in January 1957.

The British named other sites after their physical attributes. Loudwater Cove was so named because of the "thundering noise" of the sea that beats into the cove, while Breaker Island produced breakers when the sea was rough. Humble Island was so named because it was "squeezed insignificantly" between Anvers and Litchfield islands, while Laggard Island was so named because of its far-off position on the eastern fringe of the islands. Cormorant Island, as one may surmise, was named after the birds that occupied it.

In 1964, the United States established a small station, known today as Old Palmer, close to the British station, which was destroyed by fire in 1971. Meanwhile, in 1968, the U.S. Navy had started construction on the present facility on Gamage Point, within 2 kilometers east of the old site. The National Science Foundation at the time introduced RV *Hero* to the Palmer area, where it was used for research purposes and to transport personnel; it was replaced by RV *Polar Duke* in 1984-85. American ornithologists visited the Palmer area briefly before this study began in November 1973, resulting in published accounts (Watson et al. 1971; Muller-Schwarze and Muller-Schwarze 1975) and unpublished banding records by several individuals who worked in conjunction with W. J. L. Sladen's Antarctic Bird-Banding Program.

Personnel at Palmer Station continued to name sites that were left without names by the British. Shortcut Island was named because the narrow channel between it and Anvers Island provided a "shortcut" to the main body of Biscoe Bay, while Stepping Stones provided small refuge points for coastal trips in Biscoe Bay. Limitrophe Island lay at the "limit" of normal field operations from the station. Christine Island, however, was named for the wife of D. Muller-Schwarze, while DeLaca and Lipps islands were named after American scientists based at Palmer Station. One left unnamed for quite a few years was Eichorst Island, finally so named in honor of a Chicago radio operator who regularly dispatched phone calls to personnel at Palmer Station.

From the early beginnings of this study, the seas south of Anvers Island were recognized as important feeding areas for enormous numbers of seabirds, especially during the winter months, when much of Antarctica was frozen shut (Parmelee et al. 1977a; Glass 1978). A later disclosure showed that a deep basin off the Joubin Islands existed only 22 kilometers southwest of Palmer Station. Palmer Basin, as it is known today, is the only deep-water basin (1,280 meters maximum) in an area that otherwise features shallow shelves (less than 165 meters). Teeming with krill and fish, Palmer Basin was believed to be an extremely important feeding ground for seabirds and mammals, and this was verified to some extent by the sightings of banded birds that probably originated at the Palmer study area. Although much is to be learned about the physical oceanography of the area, water flow south of Anvers Island appeared to be from the southwest: from the open ocean eastward through Palmer Basin and Bismarck Strait into Gerlache Strait, thence northward toward Dallmann Bay. Coastal bays on the eastern side of Gerlache Strait and southwest side of Dallmann Bay were believed to have high rates of primary production in summer, while in autumn they accommodated feeding concentrations of postbreeding birds, and probably their young of the year. Whales and seals favored these bays as well.

According to Quetin and Ross (1988), the region's phytoplankton has been studied intensively in the area surrounding Palmer Station, where a variety of pelagic habitats ranged from shallow meristic areas to the relatively deep Palmer Basin. Recently, Holm-Hansen et al. (1989) wrote on phytoplankton blooms near the station. Zooplankton studies were, however, conducted largely in the Gerlache Strait adjacent to Liège Island by Hopkins (1985), who identified 106 species, including 4 species of copepods that were most numerous, and 3 that dominated the zooplankton biomass. However, the pelagic biomass was dominated by larger macroplanktonic species, especially the shrimplike *Euphausia superba*, commonly called krill; this finding was in sharp contrast to that of lower latitudes, where zooplankton biomass is usually greater than macrozooplankton-micronecton.

Most of the pelagic fauna research at Palmer Station focused on the biochemistry, physiology, behavior, reproduction, and growth of krill. Research on the relationship of krill to avian predators, conducted by Obst (1985), was of particular interest to this study. Generally, krill is an indispensable food for certain polar birds.

The benthic environment at Palmer included many separate antarctic communities and a wide variety of habitats (Quetin and Ross 1988). Community composition varied from place to place because of differing exposure to waves, to ice abrasion, and to light penetration. Plants dominated to depths of about 40 meters in areas of rocky substrate, and animals dominated on soft substrate. The benthic ecology (e.g., Stout and Shabica 1970), and particularly the benthic habitats and zonations, were

studied in detail by DeLaca and Lipps (1976). The area's benthic flora was particularly diverse, including approximately ninety species of marine algae, some of which thrived in protected intertidal pools and shallow subtidal regions. Brown and red algae played important roles in the area's biological system and included explosive seasonal types as well as massive bed-forming perennial types. These and other benthic forms provided substrates, shelter, and food for some of the benthic fauna on which birds fed. Seaweeds provided shags and other species with requisite plant materials for nest construction, and possibly for their courtships as well.

The diverse nature of the sea floor in the Palmer area provided the animals as well as the plants with a variety of habitats and, consequently, a diversity of species that over the years continue to attract a good many invertebrate taxonomists and ecologists to Palmer Station. Although the benthic fauna contained a variety of invertebrate species on which the area's birds fed, the one that figured prominently in this study was the Antarctic Limpet (*Nacella concinna*). It proved to be a key species in W. R. Fraser's (1989) doctoral dissertation on the Kelp Gull.

An abundant and diverse fish fauna has been described for the vicinity of Palmer

Station, including cods (Nototheniidae), ice fishes (Channichthyidae), dragon fishes (Bathydraconidae), and plunder fishes (Harpagiferidae) that are cold-adapted and live at an annual mean bottom temperature of -1.0°C. Most of the fish research carried on at Palmer Station dealt with the biochemical and physiological basis of cold adaptation and ecological aspects of feeding, reproduction, and particularly growth. Daniels and Lipps (1982) published on the distribution of these fishes as well. Most of the laboratory specimens utilized in these studies were caught far beyond Palmer at highly fertile fishing grounds near Brabant and Low islands.

The widely distributed Antarctic Silverfish (*Pleuragramma antarcticum*) figured prominently in this study. Under proper conditions, great numbers of these minnow-sized, midwater notothenoids were available to surface-feeding birds that depended upon them during their chick-rearing stages. Eastman (1985) believed that the silverfish in the Ross Sea area of Antarctica fed almost exclusively in the pelagic zone—reason enough to account for the large numbers observed at times in Palmer Basin. He also believed that this fish was also an extremely important source of

U.S. Palmer Station on Gamage Point, Anvers Island. In the foreground a Leopard Seal investigates a Crabeater Seal on an ice floe in Arthur Harbor. Photographed 20 January 1975.

food for fish as well as birds and other vertebrates; conceivably it even substituted for Euphausiids when this group was scarce. Although the silverfish's role in the polar food web may well be widely underestimated, none could deny its importance with respect to Palmer's birds.

Several species of seals occurred commonly at times in the Palmer study area. In addition to the Leopard Seal mentioned earlier, the Crabeater Seal (*Lobodon car-*

cinophagus), Southern Elephant Seal (*Mirounga leonina*), and Weddell Seal (*Leptonychotes wedelli*) pupped there. The very rare Ross Seal (*Ommatophoca rossi*) was noted near Hermit Island by several observers on 24 October 1975. Adult and immature Antarctic Fur Seals (*Arctocephalus gazella*) have increased substantially at Palmer since the mid-1970s, and, to a lesser extent, elephant seals have as well. Both seals are prone to haul out on land, and the area's vegetation suffered accordingly.

Entire moss banks were obliterated. Despite its abundance at times, the Leopard Seal did not appear to be a major predator of birds. For reasons not fully understood, seal carcasses were not eaten by many of Palmer's birds.

Whales were rarely seen within or near the Palmer study area during the 1970s and early 1980s. Many more have been sighted since that time, as they have been elsewhere in the region, possibly due to reduced whaling operations. Those species most often observed were Minke Whale (*Balaenoptera acutorostrata*), Humpback Whale (*Megaptera novaeangliae*), and the Great Killer Whale (*Orcinus orca*).

The Marr Ice Piedmont that covers nearly all of Anvers Island has receded noticeably since the beginning of this study. Alterations in the landscape were especially visible at Bonaparte Point, which may have been mostly ice covered only a hundred years ago. Exposed rock at the receding glacial edge was angular and sharp, also devoid of vegetation. As one proceeded to the tip of the peninsula, the rocks became smoother, even polished in places, and covered by a remarkable succession of plants. Although most of the ice-free surfaces of the peninsulas and outlying islands were blanketed with snow in winter, much of this disappeared by midsummer. Persistent, year-round banks of snow occurred in a few places, notably on the eastern side of Hermit Island. Meltwater rivulets occurred near the glaciers; meltwater pools both shallow and deep were everywhere evident, and although many were highly oligotrophic, others contained algae and crustaceans. Flocks of skuas, commonly called clubs, frequented meltwater pools that provided the birds with bathing opportunities and sometimes food in the form of fairy shrimp.

The study area's terrestrial flora was particularly well developed, with high species diversity and several distinct plant communities, especially on the islands long removed from glacial ice and otherwise protected from the elements. Because of its truly outstanding flora, including expansive moss banks, Litchfield Island was one of the early sites to be designated a Specially Protected Area under the Antarctic Treaty. Other areas showing unusual terrestrial flora were Norsel Point and the islands of Hermit, Shortcut, and Stepping Stones.

Although the terrestrial vegetation did not provide food directly to birds, it was used extensively by several species in nest construction. Mosses and lichens especially were important nest components, as was the grass *Deschampsia antarctica*, one of

only two vascular plants known for the region. Birds almost certainly influenced plant dispersal, most likely for one of the area's dominant lichens, *Usnea sphacellata*, and possibly for the freshwater algae that grew on soil and in runnels and ponds. Parker et al. (1972), Smith and Corner (1973), Smith (1982), and Komarkova (1983, 1984) have described the region's terrestrial flora in detail.

Terrestrial fauna was limited to four species of collembola, eleven species of mites, one tick, and one chironomid midge, so far as is known. The midge, *Belgica antarctica*, was popularly known as Palmer's flightless fly, but the arthropod that influenced the area's birds the most appeared to be the ixodid tick, *Isodes uriae*. Massive aggregations of the species were found near the Adélie Penguin colony on Torgersen Island and were studied by Lee and Baust (1982a, b, 1987). Despite the low number of species, arthropods were abundant in many snow-free areas, particularly in vegetation near bird and seal concentrations.

One of many spectacular icebergs that drift into the Palmer area and often become grounded. Photographed 26 February 1980.

C. C. Rimmer observing Antarctic Terns at a nesting cliff on Brabant Island. Photographed 30 December 1983.

anding (ringing) of the birds allowed us to identify individual adults in the study area for many years. The banding of their young provided an invaluable pool of individuals of known age and parentage that formed the basis for some of the studies (Parmelee and Rimmer 1984), and no doubt will benefit the researchers who follow.

The bands and necessary permits that accompanied them were obtained through the U.S. Fish and Wildlife Service, Office of Migratory Bird Management, Bird Banding Laboratory, Laurel, Maryland. We soon discovered that the metal bands traditionally used by the laboratory did not suit our antarctic species, which inhabited abrasive rocks, ice, and salt water; within a year the numbers on many were nearly illegible. We overcame this problem by switching to very durable Swedish-designed bands (C. G. Okman and Sons, Bankeryd, Sweden) etched deeply with numbers issued by our

laboratory. These bands proved to be so durable that they should outlive the long-lived birds on which they were placed. Nevertheless, a still better type would have been one with the numbers etched vertically on a long tube rather than horizontally; such an arrangement facilitates the reading of numbers by telescope. Although most of our subjects were easily caught repeatedly by hand or with simple hand nets, a few wary skuas and most of the gulls were not readily retrievable.

For convenient field identification of certain adults, we also used combinations of colored plastic bands, according to Bird Banding Laboratory regulations. These bands proved satisfactory for a few years, but in time some were lost and the codes then became meaningless. Moreover, the plastic bands worked down over the feet of some individuals, causing injury. During the later years of this study, we removed the plastic bands from nearly every bird retrieved.

The banded individuals made it possible to carry on studies relating to many facets of species ecology, behavior, and chronologies, including detailed field observations requisite for time-budget analyses. They also provided clues to sedentary or migratory habits. Long-distance recoveries of Palmer-banded birds were astounding.

Of particular importance to us was the retrieval of birds banded as nestlings and later returning as breeding adults to or near their natal nesting grounds. These kinds of data relate to philopatry, also called site tenacity or fidelity. Many of these polar birds require five or more years to reach breeding maturity, and peak numbers of returnees banded in this study were likely to occur in the late 1980s and early 1990s. Unless an active banding program at Palmer continues, the number of banded individuals will decline in time due to natural attrition and the inevitable disappearance of birds from the population.

For final analysis the banding data obtained in this study will be turned over to the Patuxent Wildlife Research Center in Laurel, Maryland, and the Bird Banding Laboratory. Based on the advice of the Migratory Bird Research staff, the data will be analyzed using the most current quantitative programs available for determining distributions, recovery rates, survival, density estimates, and so on.

In this brief account of the birds of Palmer Archipelago, I have attempted to include the highlights of our findings, written and arranged in a manner that will have meaning to both professional and non-professional observers. Professional ornithologists will know at a glance what species and facets of their makeup have received attention; the serious researcher, upon finding something of interest, will refer to the cited papers with their many tables, statistical treatments, and extensive coverage of the literature. The non-ornithologist engaged in other scientific disciplines should gain some appreciation of the region's avifauna. Transients traveling to that part of the world for the first time will find the birds and their polar adaptations fascinating.

In addition to the dissertations and published articles cited in this work, I have attempted to pull together fragmentary observations that otherwise surely would remain buried. Appropriate acknowledgments of unpublished material include names and the notation "personal communication"; also, observations made by those of us directly involved in this study are noted with initials (see Chapter 6, which lists all research team members). I intentionally used references only when they were indispensable to the text. No attempt was made to include general descriptions of plumages, since these have been covered adequately by Watson (1975) and by Harrison (1983).

Adjunct Studies

Dispersal of Living Organisms by Birds in the Antarctic

According to H. E. Schlichting (personal communication), no previous study had been conducted on the dispersal of algae and Protozoa by birds into and out of Antarctica. This being the case, I asked Robert Zink to collect some birds during his participation in the 1976 International Weddell Sea Oceanographic Expedition. These specimens, with additional ones from the vicinity of Palmer Station, were sealed in plastic bags, frozen, and later delivered to Schlichting for analysis. His findings (Schlichting et al. 1978) left little doubt that the birds transport such microorganisms on their bodies. From sample washings of fifteen specimens, nine rendered positive cultures containing nine species of algae, nine Protozoa, and twelve fungi. Three Cape Petrels (*Daption capense*), two Southern Giant Petrels, three Arctic Terns, and one South Polar Skua had transported microalgae and/or Protozoa. The latter two species migrate extensively between the two polar regions. All species of blue-green algae sampled from the Arctic Terns collected in Antarctica have been found as far north as the eighty-third parallel in the Canadian Arctic.

Polar birds probably transport other kinds of living forms as well. The means of long-range dispersal and establishment of a bipolar bearded lichen (*Usnea sphacellata*) has been controversial. One explanation suggests evolution of the species in the two polar regions; another promotes dispersal by birds along mountain chains, based in part on discoveries of a few intermediate localities in the Andes. The South Polar Skua inhabiting Palmer Archipelago is a good candidate for the second hypothesis, because the skuas migrate inland as well as along the coast. A banded juvenile that migrated from the Palmer study area was shot three months later by an Inuit in Greenland. It was not the first of its kind taken there, for an unbanded specimen collected there in 1902 remained long misidentified in a Copenhagen museum (Salomonsen 1976). Inasmuch as the bearded lichen grows profusely at both localities,

lichenologist John J. Thomson (personal communication) considered the South Polar Skua to be the "best vehicle yet" for explaining the lichen's puzzling distribution. Arctic Terns are poor candidates; while residing in Antarctica they hold close to the pack ice, far removed from the lichen-covered ridges where skuas raise their young.

Internal dispersal is more difficult to assess. The only incident experienced in this study occurred in February 1979, when adult Brown Skuas died mysteriously in the presence of their young, which appeared healthy. Carcasses of the adults examined at the University of Minnesota Veterinary Diagnostic Laboratories disclosed that the birds had died from a virulent form of fowl cholera, a dreadful disease of adults, but not of very young birds. Evidently this was one of very few, if not the first, documented cases of the disease in Antarctica (see the section on the Brown Skua in Chapter 17).

In this study only the initial step was taken in studying dispersal into and out of Antarctica, but the few observations indicate a fertile field for investigation. Significant advances will be made when sufficient numbers of the long-distance migrants are tracked and monitored from one hemisphere to another by satellites.

Satellite Tracking

Mapping the migratory routes of species through banding alone has limitations. Chance recoveries and their subsequent reporting not only yield low numbers of returns but also reveal only two locations or points along a given flight—the banding site and the recovery site, which may be enormously far apart. Tracking by satellite gives several points daily for many days; conceivably the transmitters of the future will last the life of the bird. Only then will we gain substantive insight into the long life and wide-ranging travels of seabirds.

Experimental bird-borne, solar-powered transmitters used in conjunction with satellites were first used on eagles and swans in North America by the U.S. Fish and Wildlife Service. Their application to seabirds appeared so promising that I traveled to Johns Hopkins University in Maryland with the hope of convincing the scientists and engineers of the Applied Physics Laboratory of a splendid opportunity: our Palmer-based giant petrels would be excellent subjects for their newly designed 200-gram transmitters. Not long thereafter Dr. Mark R. Fuller, a former University of Minnesota graduate, joined my wife and me at Palmer Station, where we experimented with harnesses and dummy transmitters for nearly a month. During the last days of January 1985 we fitted six male giant petrels with the Johns Hopkins transmitters while the big birds brooded chicks on Humble Island. The transmitters functioned well for two months on and off the breeding ground. Their signals were received by polar-orbiting *ARGOS/TIROS* satellites, and the data transmitted by way of France to the Patuxent Wildlife Research Center in Laurel, Maryland (see the section on the Southern Giant Petrel in Chapter 11).

Our satellite tracking of a giant petrel was the first ever recorded for a seabird, so far as we know. Although the use of satellite tracking of birds is still in its infancy, its potential is enormous. With such tracking it is possible not only to map move-ments and migrations, but also to learn more about behavioral and physiological traits of different species, and to develop management strategies.

Scientific Specimens

The emphasis in this study was to band and observe birds; consequently, we refrained from collecting the subjects we set out to study, unless there was a special reason to do so. However, a good many birds died, seemingly from natural causes; these for the most part were salvaged by us as research-type skeletons and flat skins. Over time, specimen numbers of certain species grew considerably; for example, more than a hundred salvageable skuas have been pre-served to date. Certain specimens were intentionally collected apart from study areas for projects involved with food, weight, molt, and so on. Winter specimens from Antarctica proved to be especially desirable.

These specimens were placed in the Bell Museum of Natural History at the University of Minnesota, Minneapolis, the National Museum in Washington, D.C., the National Museum of Natural Sciences in Ottawa, Ontario, Carnegie Museum of Natural History in Pittsburgh, Pennsylvania, and Museum of Natural History, University of Nevada, Las Vegas. One specimen of South Polar Skua was placed in the Icelandic Museum of Natural History in Reykjavik, since the opportunity exists for the sighting of the species in that part of the world.

**Satellite tracking of
Southern Giant
Petrel.**

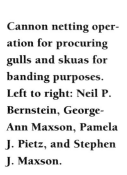

Cannon netting operation for procuring gulls and skuas for banding purposes. Left to right: Neil P. Bernstein, George-Ann Maxson, Pamela J. Pietz, and Stephen J. Maxson.

niversity of Minnesota student training was an integral part of the long-term goal of this study, which is clearly indicated by the number of graduate student participants (seven), their advanced degrees (three Ph.D., two M.S.), and published research papers (forty-one). Four of the graduate student researchers overwintered at Palmer Station and gained important insights into bird ecology and behavior otherwise unobtainable during the austral summer, when the majority of biologists conduct field studies in Antarctica. In addition, two postdoctoral researchers, several visiting scientists, undergraduate students, and field assistants contributed much to the team effort. A brief list of the participants and their itineraries is presented below in chronological order. Initials following the participants' names are used extensively (in chronological order) throughout the text to separate unpublished records from cited publications. Material not so designated is the sole responsibility of the author.

Highlights and combined time periods for each expedition are condensed for the convenience of the reader. More detailed descriptions are presented in the Appendix for all of the expedition shore landings that were conducted beyond the main study area.

Participants

David F. Parmelee (DFP) *principal investigator*. Organized and supervised all expeditions from 1972 to present.

Stuart D. MacDonald (SDM) *visiting scientist* from National Museum of Natural Sciences, Ottawa, Ontario, Canada. Collaborated with D. F. Parmelee in conducting at-sea observations in Ross, Amundsen, Bellingshausen, and Weddell seas from 24 December 1972 to 8 March 1973.

Stephen J. Maxson (SJM) University of Minnesota *graduate student* 1973-74, and *postdoctoral researcher* 1978-80. At Palmer

Station, concentrated on Antarctic Terns from 22 November to 12 December 1973, on Antarctic Blue-eyed Shags and Kelp Gulls from 12 December 1978 to 6 February 1979, and from 11 November 1979 to 12 March 1980.

William R. Fraser (WRF) University of Minnesota *graduate student*. Overwintered at Palmer Station and concentrated on Kelp Gulls from 11 January to 25 February 1975, and from 14 October 1975 to 14 October 1976. Returned to Palmer Station on another sponsored program during 1988-91, but continued observations relating to this study. Currently a senior scientist at Palmer Station on a Long Term Ecological Research (LTER) project.

David R. Neilson (DRN) University of Minnesota *graduate student*. Overwintered at Palmer Station and concentrated on South Polar Skuas and Brown Skuas from 11 January 1975 to 30 November 1975, from 8 January to 9 March 1977, and from 30 November 1977 to 22 February 1978.

Robert M. Zink (RMZ) University of Minnesota *undergraduate student*. Conducted at-sea observation in Ross, Amundsen, Bellingshausen, and Weddell seas from 16 January to 29 February 1976.

Brian M. Glass (BMG) University of Minnesota *graduate student*. Overwintered at Palmer Station and concentrated on Southern Giant Petrels from 8 January to 10 December 1977.

Neil P. Bernstein (NPB) University of Minnesota *graduate student*. Overwintered at Palmer Station and concentrated on Antarctic Blue-eyed Shags and Kelp Gulls from 26 December 1977 to 22 February 1978, and from 15 January 1979 to 12 March 1980.

William Leslie (WL) *visiting professional photographer* from Minneapolis, Minnesota. At Palmer Station, concentrated on motion pictures from 15 January to 6 February 1979, after which time he became a crew member on RV *Hero*.

Pamela J. Pietz (PJP) University of Minnesota *graduate student* 1979-81, and *post-doctoral researcher* 1985-87. At Palmer Station, concentrated on South Polar Skuas and Brown Skuas from 11 November 1979 to 12 March 1980, and from 28 December 1980 to 1 April 1981. Conducted at-sea observations in Palmer Archipelago and adjacent areas from 22 August to 22 September 1985.

George-Ann Maxson (G-AM) *field assistant*. At Palmer Station, collaborated with P. J. Pietz on South Polar Skua and Brown Skua research from 11 November 1979 to 12 March 1980.

Christopher C. Rimmer (CCR) University of Minnesota *graduate student*. At Palmer Station, concentrated on banding program involving multiple species from 23 December 1983 to 7 March 1984.

Jean M. Parmelee (JMP) *field assistant*. At Palmer Station, assisted D. F. Parmelee with banding program involving multiple species from 11 December 1984 to 30 January 1985. Participated on all MV *Illiria* and *Polar Circle* cruises.

Mark R. Fuller (MRF) *visiting scientist* from Patuxent Wildlife Research Center, Laurel, Maryland. At Palmer Station, concentrated on satellite tracking of Southern Giant Petrels from 3 January to 30 January 1985.

C. S. Strong (CSS) *visiting scientist* from Point Reyes Bird Observatory, Stinson Beach, California. Collaborated with P. J. Pietz in conducting at-sea observations in Palmer Archipelago and adjacent areas from 22 August to 22 September 1985.

K. S. Winker (KSW) University of Minnesota *graduate student*. Participated in *Polar Circle* cruises from 14 January to 23 February 1991.

Expeditions

• Expedition I 1972-73 (24 December 1972-8 March 1973). USCGC *Glacier* cruise to Ross, Amundsen, Bellingshausen, and Weddell seas by D. F. Parmelee and S. D. MacDonald.

• Expedition II 1973 (11 November-20 December 1973). Principal study area identified in the vicinity of U.S. Palmer Station by D. F. Parmelee and S. J. Maxson following RV *Hero* cruise to South Shetland Islands, Palmer Archipelago, and Antarctic Peninsula.

• Expedition III 1974-75 (26 October 1974-9 December 1975). Specific studies identified and initiated at Palmer study area by D. F. Parmelee, W. R. Fraser, and D. R. Neilson. In an unsuccessful attempt to reach the Palmer study area on 10 November 1974, Parmelee unintentionally traveled several times to the Falkland Islands and South Georgia on the RRS *John Biscoe*, HMS *Endurance*, and MS *Lindblad Explorer* before arriving at Palmer Station on 24 December.

• Expedition IV 1975-76 (14 October 1975-8 November 1976). Research continued at Palmer study area by D. F. Parmelee and W. R. Fraser. USCGC *Glacier* cruise to Ross, Amundsen, Bellingshausen, and Weddell seas by R. M. Zink (1978, 1981a, b).

• Expedition V 1976-77 (28 October 1976-15 December 1977). Research continued at Palmer study area by D. R. Neilson and B. M. Glass. Palmer Archipelago, Falkland Islands, and South Georgia visited via RRS *John Biscoe* by D. F. Parmelee (1980).

• Expedition VI 1977-78 (15 December 1977-27 February 1978). Research continued at Palmer study area by D. F. Parmelee, D. R. Neilson, and N. P. Bernstein. RV *Hero* cruise to South Shetlands.

• Expedition VII 1978-79 (3 December 1978-16 March 1979). Research continued at Palmer study area by D. F. Parmelee, S. J. Maxson, and N. P. Bernstein. RV *Hero* circumnavigated Anvers Island.

• Expedition VIII 1979-80 (6 November 1979-12 March 1980). Research continued at Palmer study area by D. F. Parmelee, S. J. Maxson, P. J. Pietz, and G-A. Maxson.

• Expedition IX 1980-81 (20 December 1980-4 April 1981). Research continued at Palmer study area by P. J. Pietz.

• Expedition X 1981 (15 October-25 November 1981). Prince Edward Islands in the Indian Ocean visited via SA *Agulas* by D. F. Parmelee.

• Expedition XI 1983-84 (19 December 1983-7 March 1984). Research continued at Palmer study area by D. F. Parmelee and C. C. Rimmer. RV *Hero* cruises along the coasts of Anvers Island, Brabant Island, and Antarctic Peninsula.

• Expedition XII 1984-85 (2 December 1984-4 February 1985). Research continued at Palmer study area by D. F. Parmelee, J. M. Parmelee, and M. R. Fuller. Historic satellite tracking of a seabird. Coast Guard C-3 Survey Boat cruises along the coasts of Anvers, Doumer, Wiencke, Joubin, and Dream islands.

• Expedition XIII 1985 (22 August-22 September 1985). *Polar Duke* WinCruise I to South Shetland Islands, Palmer Archipelago, and Antarctic Peninsula south to the sixty-seventh parallel by P. J. Pietz and C. S. Strong.

MV *Illiria* in narrow channel between the Antarctic Peninsula's precipitous coastal mountains and Petermann Island with its mixed colony of Adélie Penguins and Antarctic Blue-eyed Shags. Photographed 28 January 1990.

included at-sea observations and brief shore landings at the South Shetland Islands, Palmer Archipelago, and Antarctic Peninsula south to the sixty-fifth parallel.

Polar Circle 1991 (14 January-23 February 1991). D. F. Parmelee, J. M. Parmelee, and K. S. Winker participated in two tour cruises that included at-sea observations and brief shore landings at the Falkland Islands, South Georgia, South Orkney Islands, South Shetland Islands, Palmer Archipelago, and the Antarctic Peninsula south to the Antarctic Circle (66°33'S).

Additional Field Observations

MV *Illiria* 1988-89 (21 December 1988-7 March 1989). D. F. Parmelee and J. M. Parmelee participated in six tour cruises that included at-sea observations and brief

shore landings at the Falkland Islands, South Shetland Islands, Palmer Archipelago, and Antarctic Peninsula south to the sixty-fifth parallel.

MV *Illiria* 1990 (6 January-12 February 1990). D. F. Parmelee and J. M. Parmelee participated in four tour cruises that

Antarctic Fur Seal at Stigant Point, South Shetland Islands, R/V *Hero* in background. Photographed 19 November 1973.

King Penguin *Aptenodytes patagonicus*
current status: rare vagrant.

Adélie Penguin *Pygoscelis adeliae*
current status: year-round resident, abundant local breeder.

Chinstrap Penguin *Pygoscelis antarctica*
current status: abundant widespread breeder, rare in winter.

Gentoo Penguin *Pygoscelis papua*
current status: year-round resident, common widespread breeder.

Rockhopper Penguin *Eudyptes chrysocome*
other: *Eudyptes crestatus*
current status: rare vagrant.

Macaroni Penguin *Eudyptes chrysolophus*
current status: uncommon vagrant, probably rare local breeder.

King Penguin

Magellanic Penguin *Spheniscus magellanicus*
current status: rare vagrant.

Wandering Albatross *Diomedea exulans*
current status: rare pelagic visitor.

Black-browed Albatross *Diomedea melanophris*
other: Black-browed Mollymawk
current status: uncommon to common pelagic visitor.

Gray-headed Albatross *Diomedea chrysostoma*
other: Gray-headed Mollymawk
current status: rare pelagic visitor.

Northern Giant Petrel *Macronectes halli*
other: Northern Giant Fulmar
current status: rare vagrant.

Southern Giant Petrel *Macronectes giganteus*
other: Southern Giant Fulmar
current status: year-round resident or migrant, common local breeder.

Southern Fulmar *Fulmarus glacialoides*
other: Antarctic/Silver-gray Fulmar
current status: common year-round resident, colonies widely scattered.

Antarctic Petrel *Thalassoica antarctica*
current status: transient, rare in summer, uncommon to abundant at other times.

Cape Petrel *Daption capense*
other: Pintado Petrel/Cape Pigeon
current status; year-round resident, uncommon breeder, abundant winter migrant.

Snow Petrel *Pagodroma nivea*
current status: year-round resident, uncommon breeder in summer, uncommon to abundant at other times.

Blue Petrel *Halobaena caerulea*
current status: rare pelagic visitor.

Wilson's Storm-Petrel *Oceanites oceanicus*
current status: summer resident, abundant widespread breeder.

Black-bellied Storm-Petrel *Fregetta tropica*
current status: rare pelagic visitor.

Antarctic Blue-eyed Shag *Phalacrocorax atriceps bransfieldensis*
other: Imperial Shag/Blue-eyed Cormorant; *Notocarbo bransfieldensis*
current status: year-round resident, abundant widespread breeder.

Black-necked Swan *Cygnus melanocoryphus*
current status: rare vagrant.

Yellow-billed Pintail *Anas georgica*
current status: uncommon vagrant.

Chiloe Wigeon *Anas sibilatrix*
other: Southern Wigeon
current status: rare vagrant.

Red Phalarope *Phalaropus fulicaria*
other: Gray Phalarope
current status: rare vagrant.

Greater Sheathbill *Chionis alba*
other: American/Snowy/Wattled Sheathbill
current status: common year-round resident and widespread breeder.

South Polar Skua *Catharacta maccormicki*
other: McCormick's/Antarctic Skua
current status: summer resident, common widespread breeder.

Brown Skua *Catharacta lonnbergi*
other: Southern/Sub-Antarctic Skua; *C. skua lonnbergi/C. antarctica lonnbergi/ C. a. madagascariensis*
current status: summer resident, uncommon local breeder.

Kelp Gull *Larus dominicanus*
other: Dominican/Southern Black-backed Gull
current status: year-round resident and migrant, abundant widespread breeder.

Antarctic Tern *Sterna vittata*
current status: year-round resident, abundant widespread breeder.

Arctic Tern *Sterna paradisaea*
current status: probably an uncommon pelagic visitor.

Hypothetical List

Emperor Penguin *Aptenodytes forsteri*
Probably a rare vagrant to Palmer Archipelago. According to A. Mazzotta (personal communication) a juvenile was recorded at Paradise Bay, Antarctic Peninsula, during 1977-78. In the South Shetland Islands one was recorded by S. W. Hardee (personal communication) at Deception Island on 21 December 1975, and one by J. Valencia

(personal communication) at Fildes Peninsula, King George Island, on 25 September 1983.

Prion *Pachytilla* sp.
Two flying together alongside MV *Illiria* were observed over mostly ice-free water by C. Remadas (personal communication) in Gerlache Strait at 64°33.5' S, 62°35' W, on 28 January 1990 and likewise by E. Carriazo (personal communication) at 64°39' S, 62°51' W, on 29 January 1990. Probably these birds were Antarctic Prions (*Pachytilla desolata*), but confirmation is needed. Zink (1981a) noted *P. desolata* over open water of the Bellingshausen Sea as far south as 68°50' S, 100°20' W.

White-chinned Petrel *Procellaria aequinoctialis*
Single individuals flying near MV *Illiria* were observed by C. Remadas (personal communication) in Bransfield Strait: one at 62°23.5' S, 58°44' W, on 18 January 1990, and one at 63°01' S, 60°16' W, on 30 January 1990. Confirmation needed for Palmer Archipelago.

Pomarine Jaeger *Stercorarius pomarinus*
Probably a rare vagrant to Palmer Archi-

Northern Giant Petrel. Stray individuals follow ships to Palmer Archipelago where the birds appear to be rare. The bird shown here was photographed by the author near South Georgia on 26 November 1974. The brown-tipped bill identifies the species.

pelago. Adults were recorded south of Anvers Island (Sladen 1954).

Parasitic Jaeger *Stercorarius parasiticus*
Possibly a rare vagrant. DFP and SDM noted one far from land at 60° S, 92° W, on 8 January 1973.

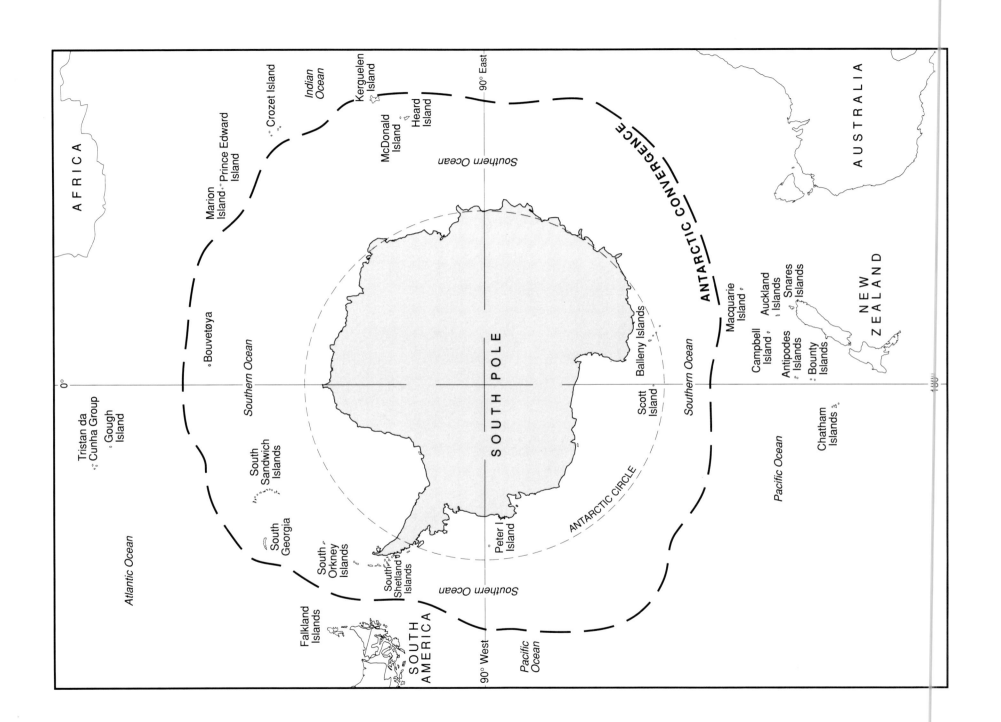

AFRICA

90° East

ANTARCTIC CONVERGENCE

AUSTRALIA

Indian Ocean

Crozet Island

Kerguelen Island

McDonald Island

Heard Island

Marion Island • ° Prince Edward Island

Southern Ocean

Bouvetøya

Southern Ocean

SOUTH POLE

NEW ZEALAND

Macquarie Island °

Auckland Islands

Snares Islands

Campbell Island °

Antipodes Islands

Bounty Islands

Balleny Islands

Scott Island

Southern Ocean

Chatham Islands

0°

Tristan da Cunha Group

Gough Island

Atlantic Ocean

South Sandwich Islands

South Georgia

South Orkney Islands

South Shetland Islands

ANTARCTIC CIRCLE

Peter I Island

Pacific Ocean

Falkland Islands

SOUTH AMERICA

90° West

Pacific Ocean

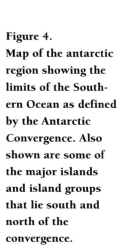

Figure 4.
Map of the antarctic region showing the limits of the Southern Ocean as defined by the Antarctic Convergence. Also shown are some of the major islands and island groups that lie south and north of the convergence.

everal antarctic life zones have been described in detail by Watson (1975), who considered the Antarctic continent and islands lying within or close to the Antarctic Circle (66°33' S) as the "Antarctic" subzone. The Maritime Antarctic subzone includes the northern Antarctic Peninsula region and southern islands of the Scotia Ridge (South Shetland, South Orkney, and most of the South Sandwich islands) and Bouvetøya (Bouvet) Island. The "Sub-Antarctic" subzone comprises the rest, but it is often subdivided further into the following subzones. The "cold" Sub-Antarctic includes extreme southern South America (Cape Horn region) and the Falkland, Prince Edward, Marion, and Crozet islands and islands south of New Zealand (Campbell, Auckland, Antipodes, Bounty). The "temperate" Sub-Antarctic includes the southern coasts of South America, Tristan da Cunha group, Gough, Amsterdam, Saint Paul, Snares, Stewart, Chatham islands, and the South Island of New Zealand. A subzone that fits somewhere between the cold

and temperate Sub-Antarctic is referred to as the "transitional" Sub-Antarctic and includes South Georgia, Kerguelen, Heard, and Macquarie islands. Observations in this study refer mostly to the Maritime Antarctic, which includes all of Palmer Archipelago.

A name used frequently today is Southern Ocean. In its broadest sense it encompasses that great body of water surrounding the Antarctic continent by combining southern portions of the Atlantic, Indian, and Pacific oceans. A more restrictive usage is to include all the ocean waters south of 60° S, thus conforming to the area covered by the Antarctic Treaty. Still another usage defines it as the ocean south of the Antarctic Convergence (Figure 4). Although the convergence is invisible to the eye, it is a well-defined boundary (detectable with thermometers) occurring wherever cold surface waters flowing northward from all around Antarctica suddenly come up against warmer and lighter waters flowing southward. The two waters do not mix readily, the heavier waters sinking below the lighter. Biologically, it is a zone of great interest because animals unaccustomed to abrupt changes in temperature succumb to stress at the convergence. If using the convergence as the northern limit of the Southern Ocean, one must be willing to view that body of water as one that also has a fluctuating boundary. The convergence is a constant line undulating between latitudes 50 and 60 degrees south, with few if any fixed points.

As pointed out by Watson (1975), the boundary lines for seabird life zones cannot be rigidly defined, because although birds are restricted to limited breeding areas in Antarctica, they are highly mobile at sea and easily transcend one or more zones, if only during feeding excursions. This is a subject that merits further study. The use of satellite tracking will help fill this void in our knowledge of seabirds.

The distribution of all these birds probably changes over time. Past records may no longer portray current known distributions, which in turn almost certainly will change as new information unfolds. In addition to my use of unpublished records, I have relied heavily on Watson (1975) and Harrison (1983) for distributional records.

Snow Petrels in Gerlache Strait.

Family of Penguins

Adélie Penguins in pack ice

O f the seven resident species of penguins listed for the Antarctic and Sub-Antarctic by Watson (1975), only three were common to Palmer Archipelago—Adélie Penguin (*Pygoscelis adeliae*), Chinstrap Penguin (*P. antarctica*), and Gentoo Penguin (*P. papua*). The latter two pygoscelids were widespread throughout the region. Breeding colonies of Adélies were largely restricted to the southwestern section of the archipelago. Their numbers increased southward along the Antarctica Peninsula, and they are considered the dominant penguin of the high latitudes all around the continent. Several penguin species typical of the Sub-Antarctic rarely penetrated Palmer Archipelago: both King Penguin (*Aptenodytes patagonicus*) and Rockhopper Penguin (*Eudyptes crestatus*) were extremely rare vagrants, while the Macaroni Penguin (*Eudyptes chrysolophus*) was an uncommon vagrant and probably rare breeder. One vagrant not recorded for the archipelago was the Emperor Penguin (*Aptenodytes forsteri*), which breeds not very far south of Anvers Island at the Dion

Islands off the southern Antarctic Peninsula. At Paradise Bay, Antarctic Peninsula, a small juvenile came ashore at the Argentine Almirante Brown Station where it was fed for five months during the 1977-78 season. According to A. Mazzotta (personal communication) it returned to sea upon completing its molt. An immature Emperor Penguin observed at the South Shetland Islands within the caldera of Deception Island by S. W. Hardee (personal communication) on 21 December 1975 almost certainly had journeyed through the archipelago, as possibly did an adult that was photographed by J. Valencia (personal communication) at Fildes Peninsula, King George Island, on 25 September 1983. Emperors wander afar: on 7 December 1975 DFP photographed an immature Emperor that came ashore near Viamonte, Tierra del Fuego, Argentina.

Adélies were the most abundant penguins along the south coast of Anvers Island, particularly in the Palmer study

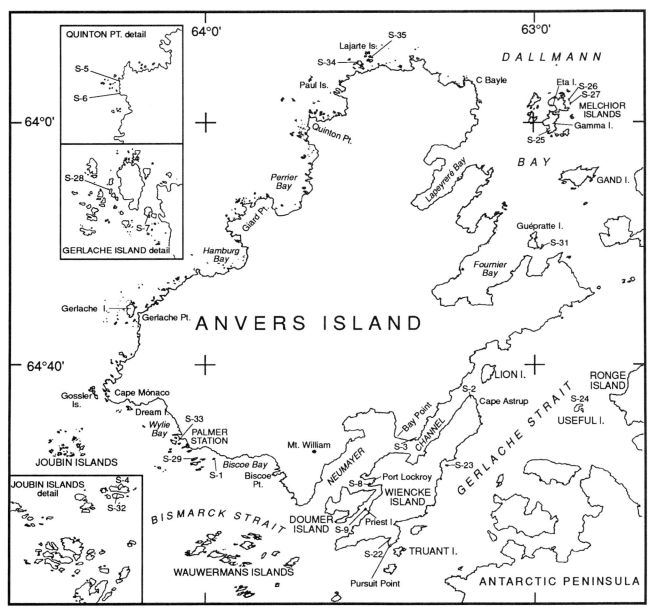

Figure 5. Approximate locations of three pygoscelid penguin colonies in Palmer Archipelago. A = Adélie Penguin (*Pygoscelis adeliae*), C = Chinstrap Penguin (*Pygoscelis antarctica*), G = Gentoo Penguin (*Pygoscelis papua*). Colonies for each species are numbered chronologically according to their published discovery dates.

area. Although this was an important species relative to our work with the skuas, no attempt was made to study it except for rough annual censuses of breeding pairs and occasional counts of chicks. The ideal time to census nesting penguins would have been during or soon following peak laying, when most pairs were present; egg attrition resulting in nest abandonment diminished the value of the counts later in the season. Although it was not always possible or feasible for us to census penguins during optimal periods, attempts were made to census at other times because mathematical techniques for handling marginal data are available to those deeply involved in penguin research.

Boating excursions beyond the Palmer study area disclosed penguin colonies that previously had not been recorded. These included large numbers of breeding Chinstraps and fair numbers of Gentoos along the west coast of Anvers Island, where Adélies proved to be scarce. Except for a single large colony of Chinstraps, Brabant Island had few breeding penguins despite its lengthy coastline. All three pygoscelids bred side by side at the Joubin Islands

(64°46' S, 64°23' W) and at a site near Gerlache Point, Anvers Island (64°36' S, 64°15' W). Revised penguin numbers and distributions for Anvers Island resulted from these censuses (Parmelee and Parmelee 1987a). While aboard their yacht *Damien II*, Poncet and Poncet (1987) mapped and estimated breeding numbers of penguins at nearly all suitable sites between 63°17' S and 69° S, compiling data on numerous unreported colonies. In combining their records with others, the Poncets produced the most complete coverage to date on penguin distribution for the region—a publication highly recommended for those interested in the subject.

Numbering colony locations has become increasingly complex. Poncet and Poncet (1987) preferred to define breeding localities by lumping several disjunct colonies within a particular locality—probably the best method for broad coverage. In restricting this study mostly to Palmer Archipelago, an island-to-island, or headland-to-headland, approach seemed appropriate, as many of the colonies near Palmer Station received individual attention. In order to simplify the numbering system within the scope of this study, each colony was numbered chronologically according to its published discovery date (Figure 5). No attempt was made to define and number

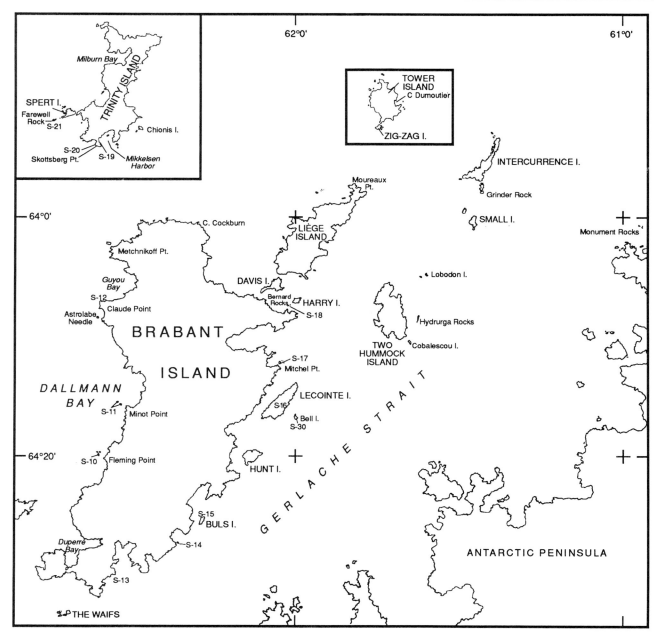

subcolonies within an island or peninsular colony. The method of Croxall and Kirkwood (1979) was followed in recording the degree of accuracy on the counts: N_1 = individual nest counts accurate to ± 5 percent (N_2 not used by Croxall and Kirkwood); N_3 = nest counts accurate to ± 10-15 percent; N_4 = accurate to 20-50 percent; N_5 = guesstimate to nearest order of magnitude (few, hundreds, thousands). Also used in these counts were the letters A (adults) and C (chicks). Poncet and Poncet (1987) used additional techniques in conducting ground counts: the letter P for the estimated numbers of breeding pairs, including corrections for nest failures. When the birds were viewed from 200 meters or more offshore,

Adélie Penguin

the number of breeding pairs (B) was estimated directly from the overall occupied surface.

Important studies dealing with distribution, feeding behavior, and other aspects of penguin research were conducted by American, Chilean, and Polish ornithologists on King George Island in the South Shetland Islands (e.g., Volkman et al. 1980; Jablonski 1984; Trivelpiece et al. 1987; Valencia et al. 1989).

Of the three species of pygoscelid penguins common to Palmer Archipelago, only the Chinstraps are near their center of abundance, with large numbers breeding there and not far away at the South Orkney and South Shetland islands and the northern half of the Antarctic Peninsula. The center of abundance for Adélie Penguins lies farther south in Antarctica, whereas that of the Gentoo Penguins is found in lower latitudes usually associated with the Sub-Antarctic. Time and again it has been noted that where the three species breed in proximity, their breeding chronologies differ significantly. The majority of Adélie Penguins are the first to lay eggs, followed by the Gentoos about two weeks later, and about two weeks after that by the Chinstraps. The only exception to the rule noted in this study was at a mixed colony near Gerlache

Point, Anvers Island, where Gentoo nestings on 3 February 1979 appeared to be somewhat behind those of the Chinstraps.

An attractive explanation for these chronological differences is given by Trivelpiece and Trivelpiece (1989) who have studied the three species for many years at the South Shetlands, principally at Admiralty Bay, King George Island. They conclude that the Chinstrap's strong nest-site bond is a product of the region's highly variable conditions, which range from warm and snow-free to ice-bound and heavily snow-covered—the very characteristics that exist at the center of the species' abundance. In addition to strong nest-site bonds, these penguins also have strong pair-bonds: a male Chinstrap will wait at its traditional nest site to re-pair with the female of a previous season—a luxury not afforded Adélie males, the majority of which nest during shorter and harsher seasons at the higher latitudes, where there is little time to wait for a mate, hence the weak pair-bond in that species. Gentoo Penguins have adapted to a more moderate climate with plentiful breeding sites; although they display a high mate-bond, they frequently change nest-site locations and, unlike their congeners, replace lost clutches. The validity of the hypothesis rests on whether Adélie and Gentoo penguins breeding at the extremity

Torgersen Island in Arthur Harbor. More than eight thousand pairs of Adélie Penguins breed annually in scattered groups in the dark, snow-free areas. Individual penguins appear as tiny specks; they are most visible against the snow. Photographed from a high-flying helicopter on 8 December 1984, several weeks after the onset of egg-laying in mid-November.

of their ranges do in fact retain the traditional behaviors of their kind that breed at the centers of abundance.

Trivelpiece and Trivelpiece (1989) also determined major differences in the foraging capabilities and strategies of the pygoscelid penguins. Conceivably the staggered breeding times reduce heavy competition for krill, the chief food consumed by these birds during demanding chick-rearing periods. Long relief periods at Adélie nests give that species the most time to search for krill, as reflected in its potential feeding range of upward of 48 kilometers—an advantage somewhat greater than that afforded Chinstraps. Gentoo Penguins are capable of diving considerably deeper than their two congeners and, consequently, by exploiting krill swarms at the deeper depths close to the breeding grounds, they are able to spend more time at their nests attending to the chicks.

King Penguin
(Aptenodytes patagonicus)

Overall Breeding Distribution. Circumpolar. Breeds in the transitional and cold subzones of the Sub-Antarctic. Subspecies *A. p. patagonicus* breeds on South Georgia and sparingly in the Falkland Islands, possibly also near Cape Horn. *A. p. halli* breeds on Prince Edward, Marion, Crozet, Kerguelen, Heard, and Macquarie islands.

Current Status in Palmer Archipelago. Rare vagrant: one record. The unusual penguin seen by Palmer Station personnel at Bonaparte Point on 20 February 1984 probably was the immature King Penguin recorded by CCR the following day at Humble Island. It was not seen again on days following.

Adélie Penguin

(Pygoscelis adeliae)

Overall Breeding Distribution. Circumpolar. Breeds chiefly within the Antarctic Circle (66°33' S) on headlands and islands around the continent, and in the Maritime Antarctic, including Bouvet Island.

Current Status in Palmer Archipelago. Adélie Penguins are year-round residents that may be encountered almost any time in Palmer Archipelago. Breeding pairs were concentrated along the south coast of Anvers Island, where they had been studied by physiologists, notably Murrish (1982), Halpryn et al. (1982), Sickles and Murrish (1983), Stemmler et al. (1984), Nagy et al. (1984), Chappell and Souza (1987), and Herwig et al. (in press). Elsewhere in Palmer Archipelago, the species was observed occasionally, usually on ice floes,

invariably in small numbers. Fledged birds of the year were uncommon, and the whereabouts of the majority of them remained unknown.

Adélie Penguin Colonies. Readily accessible colonies in the vicinity of Palmer Station have been visited by various observers since the mid-1950s. Outlying colonies have received far less attention, particularly along the west coast of Anvers Island, where the species was not known to breed before this study. According to fairly recent population estimates by Poncet and Poncet (1987), Palmer Archipelago had 36,600 breeding pairs of Adélie Penguins. Figure 4 shows the location of known colonies on or near Anvers Island. Not observed in this study were questionable colonies located at Wiencke, Hermit, and Halfway islands, and at Cape Monaco on Anvers Island (see Croxall and Kirkwood 1979).

With respect to the numbering systems used in other studies, some clarification is needed. Poncet and Poncet (1987) used single numbers for several colonies that were located fairly close to one another: No. 36 refers to Biscoe Bay, No. 37 to all the colonies near Palmer Station, No. 38 to all colonies at Joubin Islands, and No. 39 to Dream Island and a colony near by. Par-

melee and Parmelee (1987a) used Croxall and Kirkwood numbers when appropriate for newly discovered colonies the following were used: No. W-1 refers to No. A-9 this study; Nos. J-2, J-3, J-4, J-5 to Nos. A-10, 11, 12, 13; and No. G-1 to No. A-14. A chronological, abbreviated history of these colonies follows:

A-1 Torgersen Island (64°46' S, 64°05' W), Arthur Harbor. Wylie (1958) estimated between 8,000 and 10,000 nests (N_4) on 23 December 1955. Eleven nest counts were taken from 1971-72 through 1986-87, including those by members of this study, Muller-Schwarze and Muller-Schwarze (1975), Muller-Schwarze (1984), and Heimark and Heimark (1984, 1988). Except for a drastic, rather puzzling drop in the counts of Muller-Schwarze (1984) during 1983 (low of 5,523 nests [N_1] recorded), the colony has fluctuated only from a low of 8,483 nests in 1985-86 to a high of 8,876 in 1978-79, based on six counts (N_1) where nest count accuracy was judged to be ±5 percent. Considering the size and complexity of the Torgersen Island Colony, an accuracy (N_3) of ±10-15 percent is more realistic, according to Ainley and Sanders (1988). The colony showed little decline due to active tourism on Torgersen Island since the early 1970s.

Remaining to be seen is whether the 28 January 1989 oil spill in Arthur Harbor affected the penguins significantly. (Croxall and Kirkwood colony No. 25)

A-2 Humble Island (64°46' S, 64°06' W), Arthur Harbor. Wylie (1958) estimated 3,000+ nests (N_4) on 23 December 1955. Muller-Schwarze and Muller-Schwarze (1975) estimated 3,215 nests (N_3) on 2 December 1971. Ten nest counts were taken from 1974-75 through 1986-87, including those by members of this study and Heimark and Heimark (1984, 1988). With respect to seven (N_1) counts, colony size varied from a low of 2,140 in 1980-81 to a high of 2,560 in 1986-87. (Croxall and Kirkwood No. 27)

A-3 Dream Island (64°44' S, 64°14' W), Wylie Bay. Wylie (1958) reported a "huge" colony on 5 January 1957. Poncet (personal communication) estimated 12,000 chicks (C_3) on 9 January 1984; DFP and JMP guesstimated 10,000+ nests (N_5) on 10 December 1984. Heimark and Heimark (1988) estimated 10,700 nests (N_4) on 12 December 1985. Also occupied by nesting Chinstrap Penguins. (Croxall and Kirkwood No. 29)

A-4 Litchfield Island (64°46' S, 64°06' W), Arthur Harbor. Wylie (1958) estimated 1,000 nests (N_4) on 23 December 1955. Muller-Schwarze and Muller-Schwarze (1975) estimated 890 nests (N_3) on 2 December 1971. Ten nest counts were taken from 1974-75 through 1986-87, including those by members of this study and Heimark and Heimark (1984, 1988). With respect to nine (N_1) counts, colony size varied from a low of 482 in 1983-84 to highs of 650 in 1977-78 and 1980-81. The Litchfield Island colony was considerably larger in the past, judging by unused areas that showed signs of previous occupancy. (Croxall and Kirkwood No. 26)

A-5 Christine Island (64°48' S, 64°01' W), Biscoe Bay. Muller-Schwarze and Muller-Schwarze (1975) estimated 2,170 nests (N_3) on 3 December 1971. With respect to six (N_1) counts from 1979-80 through 1986-87, including those by members of this study and Heimark and Heimark (1984, 1988), the recorded low was 1,151 in 1986-87, and highs of 1,460 and 1,459 in 1979-80 and 1985-86, respectively. (Croxall and Kirkwood No. 24)

A-6 Biscoe Point (64°49' S, 63°49' W), Biscoe Bay. Muller-Schwarze and Muller-Schwarze (1975) estimated 3,020 nests (N_3) on 10 December 1971. S. Poncet (personal communication)

estimated 3,440 chicks (C_3) on 23 January 1984. DFP and JMP recorded 2,754 nests (N_1) on 21 December 1984. (Croxall and Kirkwood No. 21; Poncet and Poncet No. 36)

A-7 Cormorant Island (64°48' S, 63°58' W), Biscoe Bay. Parmelee et al. (1977a) estimated 1,000 nests (N_4) on 28 December 1974. With respect to nine (N_1) counts during 1976-77 through 1986-88, including those by members of this study and Heimark and Heimark (1984, 1988), the recorded low was 605 in 1978-79 and the high was 905 in 1984-85. (Croxall and Kirkwood No. 22)

A-8 Nameless island (64°46' S, 64°23' W), Joubin Islands. Parmelee et al. (1977b) estimated 90 nests (N_3) on 16 January 1975. PJP recorded 408 chicks (C_1) on 14 February 1981. K. Nagy (personal communication) recorded 363 nests (N_1) on 13 January 1984. S. Poncet (personal communication) recorded 330 chicks (C_1) on 24 January 1984. Also occupied by nesting Chinstrap and Gentoo penguins. (Croxall and Kirkwood No. 31)

A-9 Nameless island in Wylie Bay, 1 kilometer southwest of Dream Island. DFP, SJM, and NPB recorded 118 chicks

(C_1) on 5 February 1979. S. Poncet (personal communication) estimated 400 nests ($N_{3/4}$) on 19 January 1984. DFP and B. Obst recorded 571 nests (N_1) on 12 December 1984. Also occupied by nesting Chinstrap Penguins that were first recorded there by Poncet in 1984.

A-10 Nameless island (approximately 64°46' S, 64°24' W), Joubin Islands. DFP and CCR estimated 100 nests (N_3) on 12 January 1984. S. Poncet (personal communication) guesstimated 100 nests (N_5) on 24 January 1984.

A-11 Nameless island west of colony A-10, Joubin Islands. DFP and CCR recorded 157 nests (N_1) on 12 January 1984.

A-12 Nameless island north of Colony A-10, Joubin Islands. DFP and CCR recorded 248 nests (N_1) on 12 January 1984. S. Poncet (personal communication) recorded 365 chicks (C_1) on 24 January 1984.

A-13 Nameless island (approximately 64°45' S, 64°23.5' W), Joubin Islands. S. Poncet (personal communication) guesstimated a "few" hundred nests (N_5) on 2 January 1985.

A-14 Nameless twin islands (64°36' S, 64°15' W) west of Gerlache Point. DFP and JMP recorded 171 nests (N_1) on 2

January 1985. Also occupied by nesting Chinstrap and Gentoo penguins. (Poncet and Poncet No. 41)

Winter Records. Fluctuating numbers of adult Adélies were observed sporadically during every month of the year in the Palmer study area. According to DRN and WRF, and Heimark and Heimark (1984), the extent of ice cover influenced their abundance. The lowest numbers recorded by them occurred from late August to early October, although in 1986 the birds disappeared as early as 21 July following the formation of unusually dense pack ice that extended from Arthur Harbor to the horizon. When present, the penguins often regrouped at the colonies and engaged in low-level sexual activity characterized by courtship displays, rock picking, and sitting on nests. Conspicuous, white-throated young of the year were not observed during winter and only on rare occasions during the following spring.

Adélie Penguins inhabited pack ice when not at the breeding colonies in winter. Of all the bird species encountered by Pietz and Strong (1986), the Adélies and Snow Petrels were the ones most strongly associated with sea ice between latitudes 64°30' and 67°30' S (WinCruise I, 22 August-22 September 1985). According to WRF, Adélie Penguin survival is linked to the presence or absence of sea ice; the birds appear to forage best in the pack, whereas Chinstrap Penguins are better adapted to the open sea.

Molt. In this study a few adult Adélies in heavy molt were noted at Palmer as early as 1 February, and many by early March. In 1983, Heimark and Heimark (1984) failed to find any molting adults before 11 February, but many had commenced molting by 2 March. Molting areas occurred both near and well apart from the nesting grounds and were invariably blanketed with shed feathers where the birds were concentrated.

Predation. The Brown Skua was the principal predator of Adélie Penguin eggs and chicks in the Palmer study area, where the more abundant South Polar Skuas and Southern Giant Petrels appeared to have little influence. The often abundant Leopard Seals appeared not to be a major threat either, although they were observed eating Adélies on rare occasions.

Breeding Biology.

Nest sites. Adélie Penguins along the coasts of Anvers Island built their nests on low-lying slopes or ridges with easy access to the sea.

Nests. The nests were shallow depressions constructed of small stones splashed with guano. Dropped feathers were observed among the stones, but nest lining was not apparent. Stone thievery occurred throughout nesting.

Density. The largest concentrations of nesting Adélies in Palmer Archipelago were on Torgersen Island in Arthur Harbor (8,000 pairs) and Dream Island in Wylie Bay (10-12,000 pairs). These and other colonies were more or less fragmented into small, discrete subcolonies containing various numbers of nests, all closely spaced yet vigorously defended. Isolated nests were rare and usually close to the nests of other penguin species.

Copulation. Evidently this behavior was highly synchronized, for Heimark and Heimark (1984) observed many copulations on Torgersen Island in 1983 as early as 23 October, but by 3 November they noted only a few, apparently aborted, attempts.

Egg-laying. The earliest recorded date for eggs was 7 November 1985; S. Dame (personal communication) estimated that only 5 percent of the penguins on Torgersen Island had laid the first egg of their two-egg clutches by that time. No eggs were seen by Heimark and Heimark (1984) at the Torgersen Island colony in 1983 on 9 November, but only a week later they esti-mated that three-quarters of the large colony were incubating eggs—indicating a highly synchronized breeding effort. A few unusually late nestings occurred. DFP noted an adult incubating viable eggs and others with fairly small downy chicks on Dream Island in 1985 as late as 23 January, when most of the Adélie chicks there were rapidly replacing their down with first-winter feathers.

Clutch size. Clutch counts were not conducted in this study, but nearly all of the many observed nests had two eggs, often with one larger than the other. Three-egg clutches were not recorded.

Hatching. Eggs noted by S. Dame (personal communication) at Torgersen Island in 1985 on 7 November were observed hatching by Heimark and Heimark (1988) on 12 December—the earliest recorded hatching date in this study. Judging by the observations of Neilsen (1983) and Heimark and Heimark (1984, 1988), most Adélie eggs hatched during the second half of December.

Guard and creche periods. According to Neilson (1983), most Adélies in the Palmer area guarded their young until mid-January, when the oldest chicks started to abandon their nests and form creches; this activity continued until mid-February, when the last of the chicks disappeared. However, DFP noted a few chicks on the Humble Island nesting ground in 1980 as late as 24 February—the latest date recorded for the Palmer area. Elsewhere, a downy chick still attended at the nest was observed by DFP south of Anvers Island at Petermann Island (65°10' S, 64°10' W) in 1989 as late as 18 February, indicating an unusually late, possibly a repeat, nesting.

Synopsis of annual cycle. Adélie Penguins were concentrated in Palmer Archipelago along the south coast of Anvers Island. Although their winter activities were influenced by ice conditions, some returned from their pack-ice existence at various times to their traditional breeding sites, where they engaged in low-level sexual behavior. Large numbers returned in October and bred synchronously. Egg-laying peaked during mid- to late November; hatching took place about five weeks later. Brown Skuas were the principal predators of eggs and chicks. Following a guard period, the young abandoned their nests and formed creches. The last of the chicks left the nesting areas by mid-February, a few exceptionally later. The annual cycle from the first recorded egg (7 November) to the last recorded fledgling (24 February) was 109 days. It was not determined where the young spent the winter. Adult molt commenced in early February but did not peak until about a month later. Although Adélie Penguin foraging success and possibly survival depended on the presence of sea ice, no evidence was found that the species was declining in the Palmer area due to this or other causes, including tourism. The impact of the 1989 oil spill in Arthur Harbor was not determined.

Chinstrap Penguin
(Pygoscelis antarctica)

Overall Breeding Distribution. Circumpolar. Breeds chiefly within the American sector, including the Antarctic Peninsula and adjacent islands, South Shetland, South Orkney, South Sandwich, and South Georgia islands; also breeds sparingly around the continent on Bouvet, Heard, Balleny, and Peter islands.

Current Status in Palmer Archipelago. Chinstrap Penguins appeared to be the pygoscelid best adapted to the region, judging by the number of breeding colonies in Palmer Archipelago and adjacent areas. They were scarce along the south coast of Anvers Island, where Adélie Penguins predominated and where most bird studies had taken place. As a consequence, little is

known about the species for the archipelago, although it has been studied in the South Shetland and South Orkney islands. Winter records were scarce for the archipelago; presumably the birds migrated northward to open water beyond the winter pack.

Chinstrap Penguin Colonies. An attempt was made to record penguin colonies throughout southern Palmer Archipelago when it became apparent that Chinstrap Penguins predominated along the west coast of Anvers Island, where previously they were not known to breed. Poncet and Poncet (1987) later recorded colonies throughout the northern Palmer Archipelago. According to their population estimates, the archipelago had 26,252 breeding pairs of Chinstrap Penguins. Of this total, 7,127 pairs occupied the extreme northern section, including Tower and Trinity islands and Monument Rocks; 7,280 pairs occupied the southern extremity from Dream and the Joubin islands northward along the west coast of Anvers Island; 11,845 pairs occupied the region in between, including the largest single concentration (5,000 pairs) at Metchnikoff Point, Brabant Island. Not located by Poncets or by members of this study were Chin-

strap colonies indicated by Croxall and Kirkwood (1979) for the Melchior Islands and Auguste Island. Also not included are isolated nestings by single pairs, as have occurred at Arthur Harbor (Muller-Schwarze and Muller-Schwarz 1975; Heimark and Heimark 1988) and elsewhere. A chronological, abbreviated history of these Chinstrap colonies follows:

C-1 Hunt Island (64°20' S, 62°07' W), Gerlache Strait. Gain (1914) likely referred to this island when recording an "enormous" colony more than 100 meters high on an island south of Cape Kaiser on 27 November 1909. DFP guesstimated 200 to 300 breeding pairs (B_3) on 5 January 1978. Poncet and Poncet (1987) estimated 400 breeding pairs (B_4) on 2 February 1987. (Croxall and Kirkwood colony No. 13; Poncet and Poncet No. 72)

C-2 Cobalescou Island (64°11' S, 61°40' W), Gerlache Strait. Gain (1914) observed "two rookeries" on this island on 27 November 1909. Muller-Schwarz and Muller-Schwarz (1975) estimated 736 nests (N_3) on 16 December 1971. Poncet and Poncet (1987) estimated 500 breeding pairs ($B_{4/5}$) on 29 January 1986. (Croxall and Kirkwood No. 15; Poncet and Poncet No. 77)

C-3 Bell Island (64°16' S, 61°59' W), Gerlache Strait. Bagshawe (1938) may have referred to this island when recording a "moderate sized rookery" on 9 March 1922 on an island off Cape Kaiser. B. Obst (personal communication) guesstimated 20 to 30 nests (N_5) on 22 January 1985. Poncet and Poncet (1987) estimated 25 breeding pairs (B_4) on 29 January 1986. (Croxall and Kirkwood No. 13; Poncet and Poncet No. 73)

C-4 Palaver Point (64°09' S, 61°46' W), Two Hummock Island. Bagshawe (1938) observed a "large" colony on 3 March 1922. Poncet and Poncet (1987) estimated 3,000 breeding pairs ($B_{4/5}$) on 16 March 1986. (Croxall and Kirkwood No. 14; Poncet and Poncet No. 76)

C-5 Dream Island (64°43' S, 64°15' W), Wylie Bay. Since Wylie (1958) first recorded two nests on 5 January 1957, there has been a slow but steady increase at the colony to a high of 136 nests (N_1) when last recorded by Heimark and Heimark (1988) on 29 December 1985. Also occupied by nesting Adélie Penguins. (Croxall and Kirkwood No. 2; Poncet and Poncet No. 39)

C-6 Nameless island (63°33' S, 59°53' W) northwest of Tower Island. Rumble (in Croxall and Kirkwood 1979) recorded

Chinstrap Penguin

250 nests (N_1) on 9 January 1969. Poncet and Poncet (1987) estimated 1,500 breeding pairs ($B_{3/4}$) on 21 January 1987, indicating a substantial increase in recent years. (Croxall and Kirkwood No. 24; Poncet and Poncet No. 99)

C-7 Zig Zag Island (63°38' S, 59°50' W), south of Tower Island. Bell (in Croxall and Kirkwood 1979) estimated 700 nests (N_3) on 9 January 1969. Poncet and Poncet (1987) guesstimated 1,000 breeding pairs (B_5) on 21 January 1987. (Croxall and Kirkwood No. 25; Poncet and Poncet No. 98)

C-8 Cape Dumoutier (63°35' S, 59°44' W), Tower Island. Bell (in Croxall and Kirkwood 1979) estimated 80 nests (N_3) on 9 January 1969. Poncet and Poncet (1987) recorded only two chicks (C_1) with an estimated five pairs

Chinstrap Penguin

(P_1) on 27 January 1987, indicating a substantial reduction in recent years. (Croxall and Kirkwood No. 26; Poncet and Poncet No. 100)

C-9 Nameless island (64°46' S, 64°23' W), Joubin Islands. Since Parmelee et al. (1977b) first recorded 35 nests (N_1) on 16 January 1975, there has been a decline, with numbers fluctuating between 17 and 23 nests, based on four N_1 counts by this study, S. Poncet (personal communication), and K. Nagy (personal communication). Also occupied by nesting Adélie and Gentoo penguins. (Croxall and Kirkwood No. 1; Poncet and Poncet No. 38)

C-10 a, b, c, d Nameless islands and peninsula near Quinton Point (64°19' S, 63°41' W), Anvers Island. DFP, SJM, and NPB guesstimated a total of 8,000 breeding adults (A_5) for several disjunct colonies on 3 February 1979. Observed briefly by DFP and CCR on 27 December 1983, when a similar number was guesstimated. S. Poncet (personal communication) visited the colonies on 4 January 1990 and estimated 3,000 pairs ($P_{3/4}$) for C-10a, 2,000 pairs ($P_{3/4}$) for C-10b, and 100 nests ($N_{3/4}$) for C-10c; she also found a previously undescribed colony with an

estimated 3,500 pairs ($P_{3/4}$) at C-10d. Colony C-10a also occupied by nesting Gentoo Penguins. (Poncet and Poncet No. 42)

C-11 a, b Nameless island (64°36' S, 64°15' W) and peninsula south of Gerlache Point, Anvers Island. DFP, SJM, and NPB guesstimated 4,000 breeding adults (B_5) on 3 February 1979 for an area that included several disjunct colonies referred to in C-12 below. The colony has since been defined as two tied islands (joined at low tide) adjacent to and including a small, nameless peninsula of Anvers Island. DFP and JMP guesstimated 1,500 nests (N_5) on 2 January 1985. Also occupied by nesting Adélie and Gentoo penguins. (Poncet and Poncet No. 41)

C-12 a, b, c Nameless islands (near C-11) west of Gerlache Island. First recorded by DFP, SJM, and NPB on 3 February 1979 when combined with C-11 above. DFP and JMP guesstimated 1,000 nests (N_5) on 2 January 1985 for the three disjunct colonies on separate islands (C-12 a, b, c). S. Poncet (personal communication) estimated 86 chicks (C_3), 219 chicks (C_3), and 1,300 chicks (C_3) on 8 February 1987 for the three islands, respectively. (Poncet and Poncet No. 41)

C-13 Claude Point (64°06' S, 62°37' W), Guyou Bay, Brabant Island. Parmelee and Rimmer (1985) recorded 167 nests (N_1) on 29 December 1983. J. R. Furse (in Poncet and Poncet 1987) recorded 190 nests (N_1) on 7 January 1985. (Poncet and Poncet No. 74)

C-14 Metchnikoff Point (64°03' S, 62°34' W), Brabant Island. Parmelee and Rimmer (1985) guesstimated 5,000 nests (N_5) on 29 December 1983. J. R. Furse (in Poncet and Poncet 1987) recorded 5,000 nests (N_1), but the date is obscure. (Poncet and Poncet No. 75)

C-15 Monument Rocks (64°01' S, 60°58' W), Orleans Strait. DFP guesstimated 500 breeding pairs (B_5), including 21 nests (N_1) on the smallest islet, on 14 January 1984. Poncet and Poncet (1987) estimated 270 breeding pairs ($B_{3/4}$) on 16 January 1984. (Poncet and Poncet No. 71)

It should be noted that Trinity Island and adjacent small islands accommodate a complex array of disjunct Chinstrap colonies that is difficult to define. According to Croxall and Kirkwood (1979), Anderson recorded a colony at Skottsberg Point (63°55' S, 60°49' W) in 1905. An inshore zodiac survey by DFP on 14 January 1984 from Skottsberg Point to Farewell Rock (63°51' S, 60°55' W) disclosed at least 20 disjunct colonies ranging in size from 12 to upward of 500 pairs. These and other colonies were later recorded in greater detail by Poncet and Poncet (1987).

C-16 Nameless islet (64°53.5' S, 60°41' W) 2 kilometers southwest of Chionis Island, east of Trinity Island. Poncet and Poncet (1987) recorded two breeding pairs (B_1) on 16 January 1984. (Poncet and Poncet No. 83)

C-17 Tetrad Islands (63°55' S, 60°45' W), southeast of Trinity Island. Poncet and Poncet (1987) estimated 180 breeding pairs ($B_{3/4}$) for the three largest islands of the group on 16 January 1984. (Poncet and Poncet No. 84)

C-18 Useful Island (64°43' S, 62°52' W), Gerlache Strait. Poncet and Poncet (1987) estimated 100 breeding pairs ($B_{3/4}$) on 17 January 1984. (Poncet and Poncet No. 49)

C-19 Nameless island in Wylie Bay, 1 kilometer southwest of Dream Island. S. Poncet (personal communication) estimated 40 nests (N_4) on 19 January 1984 where none had been noted by DFP, SJM, and NPB on 5 February 1979. DFP and B. Obst recorded 47 nests (N_1) on 12 December 1984. Also occupied by nesting Adélie Penguins.

C-20 Farewell Rock (63°52' S, 61°01' W), west of Trinity Island. Poncet and Poncet (1987) estimated 100 breeding pairs ($B_{4/5}$) on 27 January 1986. (Poncet and Poncet No. 90)

C-21 Spert Island (63°52' S, 60°59' W) west of Trinity Island. Poncet and Poncet (1987) estimated 200 breeding pairs ($B_{3/4}$) on 27 January 1986. (Poncet and Poncet No. 91)

C-22 Nameless islets (63°52' S, 60°52' W) near Trinity Island, 6.5 kilometers east of Farewell Rock. Poncet and Poncet (1987) estimated 600 breeding pairs ($B_{4/5}$) for two islets on 29 January 1986. (Poncet and Poncet No. 89)

C-23 Hydrurga Rocks (64°09' S, 61°38' W), Gerlache Strait. Poncet and Poncet (1987) estimated 1,000 breeding pairs ($B_{3/4}$) on 29 January 1986. (Poncet and Poncet No. 78)

C-24 Nameless islet (64°05' S, 61°38' W) 1 kilometer southwest of Lobodon Island, Gerlache Strait. Poncet and Poncet (1987) recorded 10 breeding pairs (B_1) on 29 January 1986. (Poncet and Poncet No. 79)

C-25 Nameless islets (64°01' S, 61°28' W) close to Small Island, Gerlache Strait. Poncet and Poncet (1987) estimated 1,000 breeding pairs ($B_{4/5}$) on 29 January 1986. (Poncet and Poncet No. 80)

C-26 Nameless islet (63°58' S, 61°26' W) near Grinder Rock. Poncet and Poncet (1987) estimated 120 breeding pairs ($B_{4/5}$) on 29 January 1986. (Poncet and Poncet No. 81)

C-27 Nameless islands (63°53' S, 61°24' W) 1 kilometer northwest of Intercurrence Island, Gerlache Strait. Poncet and Poncet (1987) estimated 500 breeding pairs ($B_{4/5}$) on 29 January 1986. (Poncet and Poncet No. 82)

C-28 Trinity Island (63°54.5' S, 60°53' W). Poncet and Poncet (1987) estimated 500 breeding pairs ($B_{3/4}$) on 22 January 1987. (Poncet and Poncet No. 87)

C-29 Nameless islets (63°53' S, 60°54' W) near Trinity Island, about 4.5 kilometers southeast of Spert Island. Poncet and Poncet (1987) estimated 500 breeding pairs ($B_{4/5}$) on 22 January 1987. (Poncet and Poncet No. 88)

C-30 Trinity Island (63°49.5' S, 60°50' W). Poncet and Poncet (1987) estimated 700 breeding pairs ($B_{3/4}$) on 22 January 1987. (Poncet and Poncet No. 92)

C-31 Nameless islets (63°45' S, 60°47' W) near south coast of Milburn Bay, Trinity Island. Poncet and Poncet (1987) estimated 120 breeding pairs ($B_{3/4}$) on 22 January 1987. (Poncet and Poncet No. 93)

C-32 Trinity Island (63°43.5' S, 60°47' W), north coast of Milburn Bay, and islet 1 kilometer northwest. Poncet and Poncet (1987) estimated 350 breeding pairs ($B_{3/4}$) on 22 January 1987. (Poncet and Poncet No. 94)

C-33 Trinity Island (63°42' S, 60°48' W), 3 kilometers southeast of Megaptera Island. Poncet and Poncet (1987) estimated 200 breeding pairs ($B_{3/4}$) on 22 January 1987. (Poncet and Poncet No. 95)

C-34 Megaptera Island and nameless islet close by (63°40.5' S, 60°50' W). Poncet and Poncet (1987) estimated 900 breeding pairs ($B_{3/4}$) on 22 January 1987. (Poncet and Poncet No. 96)

C-35 Lajarte Islands (64°15' S, 63°24' W), north of Anvers Island. S. Poncet (personal communication) estimated 800 nests ($N_{4/5}$) on 5 January 1990.

Winter Records. Chinstrap Penguins visited the Palmer Station area in summer, but they did not breed there except on rare occasions. From a few to as many as fifty or more individuals appeared at various times and invariably attracted the attention of even casual observers accustomed to seeing Adélie Penguins. Despite the number of year-round observers, not a single Chinstrap Penguin was recorded near the station in winter.

Unlike the Adélie Penguins, foraging Chinstraps stayed clear of the winter pack. According to Trivelpiece and Trivelpiece (1989), the Chinstraps that bred in the South Shetlands migrated northward from there for the winter. This accounts for the meager recording of the species on Win-Cruise I by Pietz and Strong (1986) between latitudes 56°30' S and 67°30' S during 22 August-22 September 1985. The only Chinstraps observed then by Pietz and Strong (unpublished notes) were a group of six porpoising in the mostly wide open waters of Gerlache Strait at 63°57' S, 60°41' W, on 27 August 1987. It was not determined whether or not appreciable numbers of Chinstraps wintered in the broad expanses of Bransfield Strait.

Molt. On the south coast of Anvers Island in 1980, several adults in heavy molt were at Biscoe Bay on 4 March and at Arthur Harbor during 3-7 March. In the South Shetland Islands, several heavily molting adults were at Deception Island as early as 15 February in 1980, and many adults, some still attending nestlings, were undergoing very heavy molt at Half Moon Island on 17 February 1989.

Predation. Greater Sheathbills (*Chionis alba*) occupied the large Chinstrap Penguin colonies visited in Palmer Archipelago, also in the South Shetland Islands and Antarctic Peninsula, where penguin egg shells were conspicuous at sheathbill nests. The Brown Skua was the Chinstrap's principal predator on Dream Island.

Breeding Biology.

Nest sites. Some nests were situated near sea level, but more often they were found on the sides and tops of steep slopes, often 100 meters or more above the sea. Trails leading to some Chinstrap sites were deeply etched in precipitous snowbanks; some of these extended for hundreds of meters from the sea to the highest colonies.

Nests. Chinstrap nests were unlined platforms of small stones, often interspersed with a few bones, feathers, and egg shell fragments, and invariably splashed with guano. Stone thievery was common.

Copulation. In this study copulation was observed only at colonies in the South Shetland Islands. The act occurred on the nest, accompanied by much bill nipping and rubbing. Earliest date recorded was 18 November 1973 (Stigant Point, King George Island); latest date, 30 December 1977 (Aspland Island).

Breeding chronology. Unlike the Adélie Penguins, Chinstraps' egg-laying periods varied considerably, not only between seasons, but also within seasons from one locality to the next, and even from one subcolony to another within the large colonies. At Brabant Island in 1983, egg-laying had only begun to peak as late as 29 December at many Metchnikoff Point subcolonies, while some chicks had already started to

Chinstrap Penguin attending chick in the Joubin Islands. Photographed 16 January 1975.

hatch well south of there at Spigot Point on the Antarctic Peninsula. Evidently conditions were quite different the following season at Metchnikoff Point, for chicks commenced hatching there as early as 26 December, according to reports received at Palmer Station from a British party encamped near the colony. Allowing for an average incubation period of thirty-seven days (Watson 1975), clutches at those nests must have been completed by 19 November—the earliest estimated egg-laying date for Palmer Archipelago.

Egg-laying at some of the smaller colonies in Palmer Archipelago appeared to be synchronized within the season. For example, of 46 nests recorded at Dream Island on 13 January 1978, 17 had heavily incubated eggs, nine had an egg and a chick, and 20 had one or two recently hatched chicks. Clutches at some of these nests likely had been completed by 7 December, with peak laying occurring soon thereafter.

The Chinstrap's propensity to nest at various heights with different snow exposures between seasons must have contributed to asynchronized laying, as almost certainly did its tight pair-bond behavior, which tends to delay nesting. Despite these constraints, the species had sufficient time to raise its young.

Clutch size. Casual observation of many nests indicated a usual clutch of two. Nests with one egg may have been incomplete considering the asynchronous layings, or may simply have reflected loss through predation. DFP noted a nest with three fairly small young at the Joubin Islands on 16 January 1974, and another with three eggs at Dream Island on 13 January 1978. It was not determined whether one or two females had laid in these nests.

Creche formation. In this study creche formation was noted only at South Shetland Island colonies by DFP, particularly at Half Moon Island, where, in 1989, many young that had long since replaced much of their down roamed freely among molting adults on 17 February; young and old were nearly indistinguishable by 28 February that year. The early stages of creche formation were noted at this same colony, with its many subcolonies, on 30 January 1990. That day the majority of young Chinstraps, many of them still retaining much of their down, were herded into discrete groups by encircling adults that kept the chicks in place with determined, repeated jabs. Each of these many groups contained a few to upward of a dozen chicks that were forced to face inward with their heads held low for indefinite periods.

Synopsis of annual cycle. Limited information indicated that Chinstrap Penguins were well adapted to the variable climatic and breeding conditions of Palmer Archipelago. With few exceptions, the species bred colonially from near sea level to elevations of more than 100 meters, subject to local and between-season changes that often resulted in asychronous egg-laying and a wide range in hatching and fledging dates. Chinstrap breeding chronologies generally lagged behind those of the other pygoscelid penguins when breeding under similar conditions. Adult molt commenced by mid-February and was well under way by early March. Departure dates were vague, and winter sightings were scarce. Presumably most Chinstraps migrated to open water beyond the Archipelago's winter pack.

Gentoo Penguin
(Pygoscelis papua)

Overall Breeding Distribution. Circumpolar. Breeds chiefly in the transitional and cold subzones of the Sub-Antarctic, but also in the Maritime Antarctic. Subspecies *P. p. papua* breeds on Staten, Falkland, South Georgia, Prince Edward, Marion, Crozet, Kerguelen, Heard, and Macquarie islands.

Gentoo Penguins

Antarctic Blue-eyed Shags

Greater Sheathbills

**Pair of light-phased South Polar Skuas on territory at Bonaparte
Point. Torgersen Island and Litchfield Island (high bluffs) in
background. Photographed 20 January 1975.**

Pair of Brown Skuas on territory on Litchfield Island.
Glaciated Antarctic Peninsula in far background.
Photographed 15 January 1975.

**David R. Neilson at nest site of Brown Skuas on Cormorant
Island. Mt. William, Anvers Island, in background.
Photographed 17 January 1975.**

Pair of Kelp Gulls on Bonaparte Point.
Photographed 9 December 1973.

Antarctic Terns

In 1990, breeding was confirmed for the Beagle Channel near Harberton, Argentina (T. Goodall, personal communication). *P. p. ellsworthii* breeds on the South Sandwich, South Orkney, and South Shetland islands, and south along the Antarctic Peninsula.

Current Status in Palmer Archipelago. Gentoo Penguin numbers decreased from the South Shetland Islands southward to the less hospitable Palmer Archipelago and the Antarctic Peninsula, where widely scattered colonies occurred south to 65°11' S at Petermann Island. Within Palmer Archipelago, the Gentoos concentrated at Wiencke and Doumer islands, where nesting pairs were monitored sporadically over the years by other observers, and in the Joubin Islands and along the west coast of Anvers Island, where the species was not known to breed before this study. Little information on its biology was obtained, since the birds nested well beyond the Palmer study area. Although midwinter records are scarce for Palmer Archipelago, the Gentoos appeared to be year-round residents.

Gentoo Penguin Colonies. According to population estimates by Poncet and Poncet (1987), Palmer Archipelago had 9,250 breeding pairs of Gentoo Penguins. The

Gentoo Penguin

greatest concentrations were from the Joubin Islands northward along the west coast of Anvers Island (4,650 pairs) and at the combined colonies of Wiencke and Doumer islands (4,600 pairs); an additional 600 pairs were recorded for Trinity Island in the northern part of the Archipelago. Not located by Poncet and Poncet (1987) or by members of this study were questionable colonies listed by Croxall and Kirkwood (1979) for the vicinity of Cape Osterrieth (Anvers Island), Melchior Islands, and Cape Kaiser (Lecointe Island). A chronological, abbreviated history of the known extant colonies for Palmer Archipelago follows:

G-1 Port Lockroy (64°50' S, 63°31' W), Wiencke Island. According to Croxall and Kirkwood (1979), L. Gain estimated 1,000 adults (A_4) during 27-28 December 1908, and G. J. Lockley estimated 700 adults (A_4) during February-November 1945. Carroll (1954) estimated 289 chicks (C_3) on 16 Feb-

Gentoo Penguin *DFParmelee.* *30 January 1989*

ruary 1955. Tinbergen (1957, 1958) recorded 594 and 690 nests (N_1), respectively, on 12-13 December 1957 and 3 December 1958. Price (1959) recorded 447 nests (N_1) on 23 December 1959. Poncet and Poncet (1987) recorded 627 nests (N_1) and 786 chicks (C_1), respectively, on 19 January 1983 and 25 January 1984. (Croxall and Kirkwood No. 8, Poncet and Poncet No. 34)

G-2 Doumer Island (64°53' S, 63°36' W). According to Croxall and Kirkwood (1979), Roberts (unpublished data) and A. M. Carroll observed the existence of these colonies. Poncet and Poncet (1987) delineated three concentrations of subcolonies: 500 meters and 1.5 kilometers northeast of Py Point, and 1.5 kilometers east of Py Point. They recorded a total of 1,030 nests (N_1) on 22 January 1983. (Croxall and Kirkwood Nos. 6 and 7, Poncet and Poncet No. 33)

G-3 Damoy Point (64°49' S, 63°32' W) to Dorian Bay (64°49' S, 63°30' W), Wiencke Island. According to Croxall and Kirkwood (1979), Gain (1914) and Roberts (unpublished data) recorded the existence of this colony with its many subcolonies. Carroll (1954) estimated 885 chicks (C_3) on 16 February

1955. Tinbergen (1957, 1958) recorded 865 and 560 nests (N_1), respectively, on 17 December 1957 and 17 January 1959. Price (1959) recorded 685 nests (N_1) on 7 December 1959. Fletcher (in Croxall and Kirkwood 1979) recorded 885 nests (N_1) on 26 December 1978. Poncet and Poncet (1987) recorded 619 and 828 nests (N_1), respectively, on 22 January 1983 and 27 January 1984, as well as 715 chicks (C_1) on 6 February 1987. (Croxall and Kirkwood No. 9, Poncet and Poncet No. 35)

G-4 Nameless island (64°46' S, 64°23' W), Joubin Islands. Parmelee et al. (1977b) recorded 54 nests (N_1) on 16 January 1975. Colony size fluctuated from low of 41 nests (N_1) recorded on 24 January 1983 by S. Poncet (personal communication) to high of 61 nests (N_1) on 31 December 1984 recorded by DFP and JMP. Poncet and Poncet (1987) later recorded 97 chicks (C_1) on 9 February 1987. Also occupied by nesting Adélie and Chinstrap penguins. (Croxall and Kirkwood No. 31, Poncet and Poncet No. 38)

G-5 Nameless peninsula (64°21' S, 63°42' W) south of Quinton Point, Anvers Island. Small numbers of nesting adults were noted by DFP, SJM, and NPB on 2 February 1979. DFP, CCR, and K. Nagy recorded 42 nests (N_1) on 27

December 1983. S. Poncet (personal communication) estimated 90 nests ($N_{3/4}$) on 4 January 1990. Also occupied by Chinstrap Penguins. (Poncet and Poncet No. 42)

G-6 Nameless twin islands (64°36' S, 64°15' W) west of Gerlache Point, Anvers Island. DFP, SJM, and NPB guesstimated 2,000 breeding adults (B_3) on 3 February 1979. DFP and JMP recorded 1,023 nests (N_1) on 2 January 1985. Also occupied by nesting Adélie and Chinstrap penguins. (Poncet and Poncet No. 41)

G-7 North of Cape Monaco (64°39.5' S, 64°16' W), Anvers Island. DFP, SJM, and NPB recorded 950 chicks (C_1) on 5 February 1979. Presumably this colony is where Poncet and Poncet (1987) estimated as many as 2,500 chicks (C_3) on 8 February 1987, indicating substantial increase. (Poncet and Poncet No. 40)

G-8 Skottsberg Point (63°55' S, 60°49' W), Trinity Island. DFP estimated 200 nests (N_3) on 14 January 1984. Poncet and Poncet estimated 300 breeding pairs (B_3) on 22 January 1987. (Poncet and Poncet No. 86)

G-9 D'Hainaut Island (63°52' S, 60°47' W), Mikkelsen Harbor, Trinity Island. Poncet and Poncet (1987) estimated 300 breeding pairs ($B_{3/4}$) on 16 January

Pair of nesting Gentoo Penguins

1984. DFP guesstimated 300-400 nests (N_5) on 16 January 1989. (Poncet and Poncet No. 85)

G-10 Truant Island (64°55' S, 63°27' W), opposite Pursuit Point, Wiencke Island. Poncet and Poncet (1987) recorded 717 chicks (C_1) on 6 February 1987. (Poncet and Poncet No. 31)

G-11 Pursuit Point (64°54' S, 63°27' W), Wiencke Island. Poncet and Poncet (1987) estimated 200 breeding pairs

($B_{3/4}$), 1 kilometer north-northwest of Pursuit Point on 6 February 1987. (Poncet and Poncet No. 32)

Winter Records. Usually small but occasionally large numbers of Gentoo Penguins were observed near Palmer Station throughout winter. According to Holdgate (1963), the species was seen fairly regularly at Arthur Harbor during 1955-57 in March, April, and May, less commonly in June, July, August, and September, and once again regularly in October and November. Neilson (in Parmelee et al. 1977b) noted small numbers in 1975 from March through July, but none from August to mid-September; thereafter, several hundred were seen regularly until 24 October, when their numbers dropped dramatically. Heimark and Heimark (1984) noted small groups from midsummer to midwinter, and again after 10 October in 1983.

During WinCruise I, Pietz and Strong (1986) recorded Gentoos between 62° S and 65° S in the winter of 1985. Within Palmer Archipelago they recorded eight in Gerlache Strait at 64°30' S, 62°30' W, on 25 August, and one at Fournier Bay, Anvers Island, on 26 August (unpublished notes). In Lemaire Channel (65°0.35' S, 63°53.7' W), near the Antarctic Peninsula, they recorded groups of 20 and 10 on 31

August. In the South Shetland Islands they estimated 700 to 850 Gentoos in three groups at Deception Island (Whalers Bay) on 19 September, and at King George Island (Admiralty Bay) they noted "many" at the nesting grounds, some occupying nest sites.

Molt. Adults in heavy molt were noted at Port Lockroy as early as 26 January in 1989, and commonly throughout February and early March at several colonies visited that season within or near Palmer Archipelago. Some molting adults attended chicks at nests, although the majority were noted at the beaches accompanied by molting young.

Predation. The Brown Skua was the principal predator of Gentoo chicks and eggs at several colonies, notably Joubin Islands (G-4), Gerlache Point (G-6), and D'Hainaut Island (G-9). South Polar Skuas also preyed on them, judging by the many egg and chick remains at a nest close to the Gentoo colony (G-7) on Anvers Island, where Brown Skuas were absent. Greater Sheathbills inhabited most of the Archipelago's Gentoo colonies, where they scavenged and preyed on eggs and chicks. According to Watson (1975), the Leopard Seal is the major predator of Gentoos at sea, but in this

study the only recorded kill of a Gentoo by that seal was near Cuverville Island, Antarctic Peninsula, on 1 March 1989.

Breeding Biology.

Nest sites. Gentoo Penguins nested colonially from near the edge of the sea to upward of 50 meters elevation on fairly steep inland ridges. They were the most enterprising pygoscelid in reclaiming traditional nesting grounds at abandoned human sites, as demonstrated by their many nests that covered the old station grounds at Port Lockroy, even the interiors of buildings at Waterboat Point, Antarctic Peninsula.

Nests. Both sexes engaged in building nests of small stones, some of the structures amounting to elaborate bowls, others hardly discernible from scrapes in the ground. The stony nests often contained old bones, feathers, egg shells, occasionally mummified young from past seasons, all copiously splashed with guano. Isolated nests were not uncommon, but most were closely spaced in tight but well-defended territories. Stone picking and thievery were common.

Breeding chronology. The few observations in this study indicated that even within seasons the species' egg-laying periods and related chronologies varied considerably from one locality to another. Egg-laying had already commenced in the South Shetland Islands (Harmony Cove, Nelson Island) on 13 November 1973, while at the Antarctic Peninsula (Paradise Bay) that same season, the Gentoos had only begun to occupy their mostly snow-covered nests on 19 November. Colonies close to one another also showed chronological differences during the same season; for example, in 1989, the majority of the Gentoos at the Port Lockroy colony (G-1) were behind those that nested not far away at Paradise Bay.

Environmental differences between seasons also influenced egg-laying at the same colony. Eggs but no young were noted at Port Lockroy nest sites as late as 29 December in 1983, whereas young fully a third grown occupied identical sites as early as 5 January in 1989. Gentoos occupying the higher, more exposed ridges generally nested and vacated their sites sooner than those at lower elevations, presumably due not only to experienced breeders, but also to local differences in snow cover. Asynchronous nestings also occurred at certain colonies where elevation and snow cover were uniform throughout; for instance, at the Gerlache Point colony (G-6) on 2 January 1985, 1,023 nests held both fresh and incubated eggs, as well as newly hatched to fairly large chicks. On the other hand, all 42 nests at the smaller Quinton Point col-

ony (G-5) held only eggs on 27 December 1983, indicating a synchronous though late laying.

A few Gentoo nestings were extremely tardy. Two pairs incubated viable eggs at the Cape Monaco colony (G-7) in 1979 as late as 5 February, when most of the many nests there had large chicks. No evidence was found that late nestings necessarily resulted in failure. Two recently hatched siblings were noted near the Antarctic Peninsula (Cuverville Island) as late as 18 February in 1989; both were well fed and robust when checked last on 1 March.

The tight pair-bond, which often requires time for previous mates to remate, and the species' propensity to replace lost clutches perhaps contributed to the Gentoo's asynchronous egg-laying and consequent extended breeding season. The chronological spacing of nestings within large colonies may well spread the krill resources during critical chick-rearing periods—an advantage analogous to that described by Trivelpiece and Trivelpiece (1989) for mixed colonies.

Clutch size. Two eggs appeared to be the usual clutch. Three-egg clutches were not recorded.

Creche formation. The earliest date recorded for Palmer Archipelago was 26 January (Port Lockroy, 1989). Unlike the Adélie Penguins, which abandoned the

breeding areas soon after creching, the Gentoos remained well into March; old and young together were strung out along the stony terraces for indefinite periods. According to Trivelpiece and Trivelpiece (1989), young Gentoos practice swimming at sea, but they return frequently to their natal areas, where they are fed by their parents.

Synopsis of annual cycle. Gentoo Penguins were year-round residents in Palmer Archipelago, though midwinter observations were scarce. New colonies were recorded, but virtually nothing was learned of the species' arrival times and early courtship behaviors. With few exceptions the majority of Gentoos bred later than the majority of Adélies, but earlier than the majority of Chinstraps, especially when nesting together. Breeding chronologies varied between seasons and locations, also within single large colonies, probably due to several factors, including local snow conditions, tight pair-bonding, and replacement clutches. Although early egg-laying dates were not obtained, the size of some chicks in early January indicated that eggs must have been laid by late November. February egg dates extended the extra-long period of seventy or more days for the occurrence of eggs. Survival at some late nestings was high. Contributing to the

Macaroni Penguin

lengthy breeding season was the long creching period, when adults and young molted together at the beaches.

Rockhopper Penguin
(Eudyptes chrysocome)

Overall Breeding Distribution. Circumpolar. Transcends all three subzones of the Sub-Antarctic. Subspecies *E. c. chrysocome* breeds near Cape Horn and the Falkland Islands; *E. c. moseleyi* at Tristan da Cunha,

Gough, St. Paul, and Amsterdam islands; and *E. c. filholi* at Prince Edward, Marion, Crozet, Kerguelen, Heard, Macquarie, Campbell, Auckland, Antipodes, and Bounty islands.

Current Status in Palmer Archipelago. Rare vagrant: two records. An adult male with a broken wing was collected within the Palmer study area at Cormorant Island on 29 December 1980, approximately 1,100 kilometers from the nearest breeding grounds for the species, the islands off Cape Horn (Matthew 1982). The specimen was later deposited at the Academy of Natural Sciences of Philadelphia. The second record was obtained on 26 December 1984 by way of a radio dispatch to Palmer Station from a British expedition team encamped at Metchnikoff Point, Brabant Island. One Rockhopper Penguin and one Macaroni Penguin had been "around for a few days," according to one of the expedition personnel under J. R. Furse.

Macaroni Penguin
(Eudyptes chrysolophus)

Overall Breeding Distribution. Breeds chiefly in the transitional and cold subzones of the Sub-Antarctic, but also in the Mar-

itime Antarctic. Subspecies *E. c. chrysolophus* breeds on the Falkland Islands, islands of the Scotia Ridge, and on Bouvet, Prince Edward, Marion, Crozet, Kerguelen, and Heard islands. The Royal Penguin, *E. c. schlegeli*, considered by some researchers to be simply a color phase, breeds on Macquarie Island.

Current Status in Palmer Archipelago. Uncommon vagrant; probably rare local breeder. Locations of some breeding colonies in the general region of the Antarctic Peninsula have been doubted in recent years (see Volkman et al. 1982). Most records for Palmer Archipelago derive from Palmer Station, where observers constantly monitor the area's wildlife. No doubt the species occurs more frequently in the Archipelago than records indicate, since small colonies exist not far away in the South Shetland Islands.

Summer Records. Within the Palmer study area, Holdgate (1963) reported a crested penguin for Humble Island on 6 January 1956, identified at first as a Rockhopper Penguin but later as a Macaroni Penguin. In 1979, Bernstein and Tirrell (1981) observed one at Cormorant Island on 21 February and later collected possibly the same individual (adult female in heavy molt) at Humble Island on 3 March. CCR, J. G. Morin, and K. Nagy observed one at Torgersen Island on 26 January 1984. DFP and JMP observed what appeared to be the same individual each time at Humble Island during 12-28 January 1985. Heimark and Heimark (1988) recorded the first documented nesting attempt for the Archipelago. At Humble Island they observed one adult with a single egg, presumably its own, during the 1985-86 season from 5 to 24 December, after which date the egg disappeared; one adult was next seen near the site on 5 January and again on 7 February, but not thereafter. The single egg was not unusual for Macaroni Penguins, for the species customarily keeps only the larger of the two eggs that it usually lays. The nest was situated at the edge of an Adélie Penguin colony.

Several sightings have been recorded beyond the Palmer study area. G. E. Watson collected a female (oviduct slightly enlarged) on Dream Island west of Palmer Station on 29 January 1966 (R. C. Banks, personal communication). According to a 26 December 1984 radio dispatch received at Palmer Station from a British expedition under J. R. Furse, one Macaroni Penguin and one Rockhopper Penguin were observed on Brabant Island at Metchnikoff Point—the site of a large Chinstrap Penguin colony. At Port Lockroy, Wiencke Island, DFP and JMP photographed a single Macaroni Penguin first observed by E. Carriazo on 7 February 1990; the bird, in extremely heavy body molt, held its ground among nesting Gentoo Penguins and nipped Gentoo chicks that came too close.

In this study the only Macaroni colonies visited in the South Shetlands were located on Elephant Island: forty nests (N$_3$), one with a newly hatched chick, the rest with single eggs (one pipped), at Walker Point on 19 December 1974; twelve breeding pairs (B$_5$) at Cape Lookout on 3 February 1989 and 26 January 1990 (DFP).

Magellanic Penguin
(Spheniscus magellanicus)

Overall Breeding Distribution. Falkland Islands, Argentina, and Chile southward to Cape Horn and vicinity.

Current Status in Palmer Archipelago. Rare vagrant: one record. WRF and CCS noted one on Dream Island west of Palmer Station on 20 February 1990. According to Watson (1975), this penguin occurs at sea and as a very rare vagrant in New Zealand, South Georgia, and possibly the Tristan da Cunha group.

Family of Albatrosses

**Gray-headed
Albatross**

Albatrosses could be observed almost any time in the windy, often turbulent Drake Passage, but their numbers dropped precipitously south of there in the icy doldrums of Palmer Archipelago and the Antarctic Peninsula, where albatross flight was hampered by lack of winds. Only the Black-browed Albatrosses (*Diomedea melanophris*) penetrated those areas with regularity, due perhaps to the species' ability to utilize modest winds, such as occur at times along the sea channels and large freshwater lakes of Argentina and Chile, where the birds are frequent visitors. Although the albatrosses bred a long way from Palmer Archipelago, the wide-open seas west of Anvers Island and Bransfield Strait attracted foraging individuals. Probably these birds were more numerous than currently acknowledged due to the paucity of observers in the little-visited waters. Since nearly all albatross records were obtained at sea, one concludes that these birds do not often come within sight of shore-based people. Unusually stormy periods may drive the birds closer to land, as apparently was the case during the summer of 1990, when on several occasions WRF saw one to three individuals flying surprisingly close to Palmer Station.

Wandering Albatross
(*Diomedea exulans*)

Overall Breeding Distribution. Circumpolar. Breeds in all three subzones of the Sub-Antarctic. Subspecies *D. e. exulans* breeds at Inaccessible, Gough, Amsterdam, Auckland, Campbell, and Antipodes islands. *D. e. chionoptera* breeds at South Georgia, Prince Patrick, Marion, Crozet, Kerguelen, and Macquarie islands.

Current Status in Palmer Archipelago. Rare pelagic visitor: one record. DFP recorded one flying over unusually rough seas in Bransfield Strait at 62°20' S, 58°23' W, on 12 January 1990. According to D. Schoeling (personal communication), this possibly was the same bird that he had seen following the ship northward to King

Black-browed Albatross

George Island in the South Shetlands. The paucity of records was surprising in view of the species, abundance in the Drake Passage. Wandering Albatrosses do reach high latitudes at times, for DFP recorded them as far south as the seventy-first parallel in the Weddell Sea, where one individual flew with labored wing beats over unbroken expanses of pack ice on 1 February 1973.

Black-browed Albatross
(*Diomedea melanophris*)

Overall Breeding Distribution. Circumpolar. Breeds in the transitional and cold Sub-Antarctic. Subspecies *D. m. melanophris* breeds chiefly on the Falkland, South Georgia, Kerguelen, Heard, Antipodes, Macquarie, and Campbell (few) islands. *D. m. impavida* also breeds on Campbell Island.

Current Status in Palmer Archipelago. Uncommon to common pelagic visitor; recorded sporadically from 9 December to 19 April. The extraordinary count by WRF off Anvers Island on 2 March 1986 merits comment. His estimated number of 750 Black-browed Albatrosses, including flocks of 120 and 300 birds, consisted of 85-90 percent juveniles scattered for a distance of about 25 kilometers. All fed above a shelf 350 meters deep. Other than that, flocks were not often encountered. DFP, SJM, and NPB noted a flock of six following a school of Crabeater Seals northwest of Anvers Island at 64°24' S, 64°17' W, on 1 February 1979. DFP noted a small feeding flock of five Black-browed Albatrosses composed of two adults and three juveniles at Dallmann Bay on 3 March 1989. Although age determinations could not always be ascertained at sea, the majority of Black-browed Albatross sightings (Table 1) were of single, isolated flying adults, though occasionally several were in view at the same time.

Gray-headed Albatross
(*Diomedea chrysostoma*)

Overall Breeding Distribution. Circumpolar. Breeds in the transitional and cold Sub-Antarctic, including Diego Ramirez off Cape Horn, South Georgia, Prince Edward, Marion, Crozet, Kerguelen, Macquarie, and Campbell islands.

Current Status in Palmer Archipelago. Rare pelagic visitor: five records, all near Anvers Island. DFP recorded one flying near the Joubin Islands on 23 January 1975. DRN recorded two flying individuals near Arthur Harbor on 27 February 1975. WRF recorded one flying near Arthur Harbor on 19 April 1976. DFP, SJM, and NPB noted at least four widely scattered individuals off the west coast of Anvers Island between the Gossler and Paul islands on 1 February 1979.

Wandering Albatross

Table 1. Sightings of Black-browed Albatross (*Diomedea melanophris*) in Palmer Archipelago

No.	Date	Locality	Observer(s)
1	23 Jan. 75	Bismarck Strait, off Joubin Islands	DFP
1	27 Feb. 75	Bismarck Strait, off Joubin Islands	DRN
1	19 Apr. 76	Bismarck Strait, off Arthur Harbor	WRF
4	15 Feb. 77	Bismarck Strait, off Arthur Harbor	DRN
1	15 Jan. 79	Gerlache Strait, off Two Hummock Island	DFP
6	01 Feb. 79	NW Anvers Island, 64°24′ S, 64°17′ W	DFP
1	02 Feb. 79	NW Anvers Island, 64°15′ S, 63°10′ W	DFP
1	03 Feb. 79	NW Anvers Island, 64°09′ S, 63°53′ W	DFP
2	16 Feb. 80	Dallmann Bay, 64°27′ S, 62°59′ W	DFP
1	16 Feb. 80	Dallmann Bay, near Melchior Islands	DFP
1	27 Dec. 83	NW Anvers Island, 64°40′ S, 64°24′ W	DFP, CCR
1	27 Dec. 83	NW Anvers Island, 64°20′ S, 64°41′ W	DFP, CCR
2	29 Dec. 83	Bransfield Strait, 64°07′ S, 62°39′ W	DFP, CCR
2	30 Dec. 83	Bransfield Strait, off Brabant Island	DFP, CCR
1	11 Jan. 84	Bismarck Strait, off Joubin Islands	B. Obst
2	12 Jan. 84	Bismarck Strait, off Joubin Islands	DFP, CCR
1	08 Feb. 84	Bismarck Strait, off Joubin Islands	CCR
1	09 Dec. 84	Dallmann Bay, off Melchior Islands	DFP
1	31 Jan. 85	Bransfield Strait, off Liège Island	DFP, JMP
1	31 Jan. 85	Bransfield Strait, off Hoseason Island	DFP, JMP
750	02 Mar. 86	SW Anvers Island, 64°52′ S, 64°48′ W	WRF
1	27 Jan. 89	Bismarck Strait, off Arthur Harbor	WRF
1[a]	29 Jan. 89	Bransfield Strait, 63°45′ S, 60°53′ W	DFP
1	01 Mar. 89	Bransfield Strait, 63°37′ S, 61°10′ W	DFP
1	01 Mar. 89	Bransfield Strait, 64°05′ S, 61°45′ W	DFP
5	02 Mar. 89	Bismarck Strait, off Biscoe Point	DFP
1	03 Mar. 89	Dallmann Bay, 64°12′ S, 63°00′ W	DFP
5	03 Mar. 89	Dallmann Bay, 64°05′ S, 63°04′ W	DFP
1	05 Feb. 90	Bransfield Strait, 63°36′ S, 61°09′ W	DFP

[a] Feeding in the water in company with a Southern Giant Petrel and two South Polar Skuas.

Family of Giant Petrels, Fulmars, Petrels, Prions, Gadfly Petrels, and Shearwaters

Petrels at sea.

Watson (1975) listed twenty-three species of procellariids for the Antarctic and Sub-Antarctic, including both resident and migrant breeders; an additional three species were treated as either nonbreeding migrants or vagrants. Four of the species from Watson's list were found breeding within the Palmer Archipelago and adjacent areas: Southern Giant Petrel (*Macronectes giganteus*), Southern Fulmar (*Fulmarus glacialoides*), Cape Petrel (*Daption capense*), and Snow Petrel (*Pagodroma nivea*). No evidence was found that Antarctic Petrels (*Thalassoica antarctica*) bred anywhere in the region, though they were regular visitors. The Northern Giant Petrel (*M. halli*) and Blue Petrel (*Halobaena caerulea*) were rare pelagic visitors, while unusual sightings of prions and petrels require confirmation.

Southern Giant Petrels, also referred to as Southern Giant Fulmars, were the only procellariids that bred within the Palmer study area, and consequently they received a great deal of attention. An attempt was made to map the breeding grounds of the other species, but such sites within the Archipelago proved to be few and far between. Most perplexing was the continued failure to find significant colonies of Southern Fulmars, since large numbers of them were sighted repeatedly at sea, particularly in the oft-traveled Gerlache Strait. Major colonies remain to be located and described.

Northern Giant Petrel
(*Macronectes halli*)

Overall Breeding Distribution. Circumpolar. Breeds in all three subzones of the Sub-Antarctic, including Gough, South Georgia, Prince Edward, Marion, Crozet, Kerguelen, Macquarie, Campbell, Auckland, Stewart, Antipodes, and Chatham islands.

Current Status in Palmer Archipelago. Rare vagrant. WRF observed a single Northern Giant Petrel near Palmer Station during

6-16 January 1976. The bird had joined a Chilean vessel some 600 kilometers north of Anvers Island and had followed the vessel all the way to Arthur Harbor, where it finally abandoned the ship. During its brief residency at Palmer it attempted to catch gull chicks—a predatory behavior not observed among the many Southern Giant Petrels residing in the area. Although a diverse diet is known for both species of giant petrels, some interspecific differences in the diets and feeding behaviors occur, especially where the two breed sympatrically at different times of the year. Johnstone (1977, 1979) suggested that *M. giganteus* had a more pelagic feeding regime than *M. halli,* but Hunter (1983) believed that intersexual differences in the diet of the two species are more marked than interspecific ones. According to Hunter, females of both species are prone to feed at sea, while the males of both species feed more on carrion, depending to a certain extent on the availability of foods at given periods. More studies are needed in areas of sympatry as well as allopatry.

Six-year-old Southern Giant Petrel. This individual was banded when a nestling at Norsel Point on 11 February 1979 and photographed when a breeding adult male at Norsel Point on 20 January 1985.

Ten-year-old Southern Giant Petrel. This individual was banded when a nestling at Shortcut Island on 9 April 1975 and photographed when a breeding adult female at Shortcut Island on 4 January 1985.

Sixteen-year-old Southern Giant Petrel. This bird was banded when a nestling at Bonaparte Point on 28 February 1975 and photographed when a breeding adult male at Bonapart Point on 9 February 1991.

Twenty-six-year old Southern Giant Petrel. Banded when a nestling at Bonaparte Point before this study on 18 February 1965 and photographed when a breeding adult female on Bonaparte Point on 9 February 1991.

Southern Giant Petrel
(*Macronectes giganteus*)

Overall Breeding Distribution. Circumpolar. Breeds in the transitional subzones of the Sub-Antarctic, and in the Maritime Antarctic and Antarctic. Includes the Falkland, Prince Edward, Marion, Crozet, Kerguelen, Heard, South Georgia, South Sandwich, South Orkney, and South Shetland islands, Antarctic Peninsula area, and Adélie Coast, Windmill Islands, and Enderby Land on the main continent.

Current Status in Palmer Archipelago. Adult Southern Giant Petrels occurred at sea throughout Palmer Archipelago, where they may be observed almost anytime year-round. Other than the South Shetland Islands, where the species breeds commonly, nesting pairs were concentrated within the archipelago only along the southern and southwestern coasts of Anvers Island. During 1986-87, S. Poncet (personal communication) observed an uncertai number of nesting pairs near the Antarctic Peninsula at Moss Islands (64°09' S, 61°03' W) and nearby Sterneck Island (64°11' S, 61°02' W). A few young of the year wintered in the vicinity of Palmer Station, but most migrated from the breeding grounds within a short time. Glass (1978,

Pair of Southern Giant Petrels at Humble Island nest site. Note the larger bill of the male at upper left. Photographed 15 December 1984.

unpublished notes) made numerous observations on the species' behavior during the winter months, while Parmelee et al. (1985), Fuller (in Strikwerda et al. 1986), and Parmelee and Parmelee (1987b) concentrated on banding and bird-borne transmitter/satellite projects. Giant petrels were popular subjects for physiological studies at Palmer Station as well, notably by Murrish and Guard (1977), Guard and Murrish (1979), Tirrell and Murrish (1979a, b), Murrish and Tirrell (1981), and Ricklefs (1982).

Table 2. Southern Giant Petrels (*Macronectes giganteus*) banded as adults (age uncertain) at their nests near Palmer Station

Season banded	Nesting site	No. males banded	No. females banded	No. unsexed banded
1974–75	Humble Island	2	3	7
	Bonaparte Point			4
	Shortcut Island		1	1
	Litchfield Island			1
1975–76	Bonaparte Point	1		
1976–77	Humble Island	8	7	
	Bonaparte Point	8	5	1
	Shortcut Island			2
1977–78	Bonaparte Point		1	2
	Litchfield Island			1
1978–79	Cormorant Island	1		
1984–85	Humble Island	15	15	
	Stepping Stones	3	1	
	Total	38	33	19

Table 3. Foreign-banded Southern Giant Petrels (*Macronectes giganteus*) recovered at their nests near Palmer Station

Band no.	Date banded	Recovery locality	Recovery date	Locality	Sex	Remarks
130-50062	15 Aug. 70	Australia, Malabar (33°58′ S, 151°20′ E)	19 Jan. 79	Litchfield Island	?	distance 8,700 km
130-60253[a]	06 Aug. 72	Australia, Victoria (38°17′ S, 144°30′ E)	05 Feb. 81	Humble Island	F	distance 8,300 km
131-22140	16 Aug. 75	Australia, NSW (28°39′ S, 153°37′ E)	11 Jan. 84	Hermit Island	M	distance 9,214.7 km
			04 Jan. 89	Hermit Island		

[a] Sacrificed for physiological research at Palmer Station.

Banding. A total of 1,936 Southern Giant Petrels were banded within the Palmer study area.

Adults. Ninety adults of uncertain age were banded (Table 2). Of these, 27 were additionally color banded and formed the basis for BMG's observations in 1977. The banded adults provided information on site and mate fidelity and absenteeism, and they were used in the 1985 satellite tracking project. To date there have been no foreign recoveries of any of these birds. Foreign-banded adults that were recovered at nests within the Palmer Study area are listed in Table 3.

Nestlings. Of the 1,846 nestlings banded, 23 were recovered at foreign sites to date (Figure 6; Tables 4, 5, 6). For comparable results of banding programs involving giant petrel nestlings at South Georgia, see Hunter (1984a).

Of special interest were the recoveries of Palmer-banded nestlings that later bred at or near their natal sites (Table 7). Of 76 nestlings recovered, 63 (82 percent) were sexed and showed nearly an equal return of males and females.

To date there have been only two returns of Palmer-banded nestlings that later bred outside the Palmer study area. No. 608-34906, banded 23 March 1977 by BMG on Norsel Point, was recovered not

Table 4. Long-distance recoveries of Southern Giant Petrels (*Macronectes giganteus*) banded as nestlings near Palmer Station

Band no.	Banding date	Recovery locality	date	Locality
0658-67539 (1)	25 Mar. 75	Stepping Stones	27 July 75	Australia, Yanchep Beach 31°30' S, 115°30' E
0658-67555 (2)	25 Mar. 75	Stepping Stones	08 July 75	Australia, Pt. Sir Isaac 34°20' S, 135°10' E
0658-67562 (3)	03 Apr. 75	Shortcut	24 Aug. 75	Australia, 3S Bald Head 35°10' S, 118°00' E
0658-67573 (4)	03 Apr. 75	Shortcut	76	Australia, NR Eagle Ridge 29°40' S, 114°50' E
0658-67590 (5)	09 Apr. 75	Shortcut	11 Oct. 75	Australia, 6 NW Rainbow Beach 27°20' S, 153°20' E
0658-67593 (6)	18 Apr. 75	Humble	20 July 75	Australia, Jandakot 32°00' S, 115°50' E
0658-67648 (7)	19 Mar. 76	Humble	14 July 76	New Zealand, Titahi Bay 41°00' S, 174°50' E
0608-34637 (8)	15 Mar. 77	Hermit	21 June 77	Australia, NR Kalbarrie 27°80' S, 114°00' E
0608-34721 (9)	15 Mar. 77	Stepping Stones	29 July 77	Australia, 10 E Eden NSW 37°05' S, 150°00' E
0608-34794 (10)	22 Mar. 77	Norsel Point	26 July 77	Australia, Coffin Bay 34°20' S, 135°10' E
0608-34864 (11)	11 Feb. 78	Stepping Stones	19 July 78	Australia, Mandurah 32°30' S, 115°40' E
0608-34860 (12)	11 Feb. 78	Stepping Stones	10 Aug. 78	Australia, Blanche Port 32°40' S, 134°10' E
1217-00101 (13)	07 Mar. 79	Stepping Stones	Nov. 79	Australia, 25 SW Fremantle 32°30' S, 115°40' E
1217-00226 (14)	07 Mar. 79	Stepping Stones	19 Sept. 80	Falklands, Pebble Island 51°20' S, 59°40' W
1217-00205 (15)	07 Mar. 79	Stepping Stones	06 Apr. 80	Australia, McGrath Flat 35°50' S, 139°10' E
1217-00262 (16)	13 Mar. 79	Humble	28 Sept. 79	Chile, PTA DeTopocalma 34°00' S, 72°10' W
1217-00440 (17)	09 Feb. 80	Humble	June 80	Australia, 7 N Wallaroo 33°40' S, 137°30' E
1217-00571 (18)	13 Feb. 80	DeLaca	16 July 80	Australia, NR Point Quobba 24°30' W, 113°20' E
0648-05329 (19)	11 Feb. 84	Shortcut	28 Aug. 84	South Africa, Cape Receife 34°00' S, 25°40' E
0648-05602 (20)	24 Feb. 84	Stepping Stones	28 June 85	Brazil, Cidreira Beach 30°10' S, 50°10' W
0648-05720 (21)	03 Apr. 87	Humble	16 July 87	Australia, 15 N Mandura 32°10' S, 115°40' E
0648-05752 (22)	12 Apr. 87	Norsel Point	07 July 87	Australia, NR Breaksea Island 35°00' S, 118°00' E
0648-05893 (23)	14 Feb. 89	Shortcut	12 Dec. 89	New Zealand, Kaikoura Peninsula 42°20' S, 173°40' E

Figure 6. Approximate recovery sites of 23 Southern Giant Petrels (*Macronectes giganteus*) banded as nestlings near Palmer Station. The effort expended by the Australians in recovering banded birds probably reflects the high recovery rate for that part of the world.

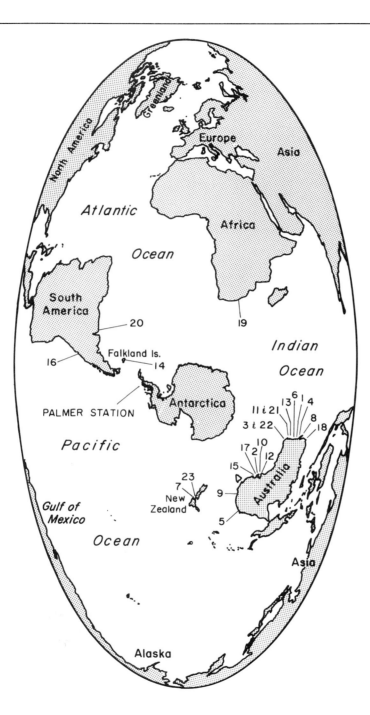

Table 5. Long-distance recovery rates of Southern Giant Petrels (*Macronectes giganteus*) banded as nestlings near Palmer Station

Year banded	No. banded	Recovered No.	(%)	Year(s) recovered
1975	105	6	(5.71)	1975(5), 1976(1)
1976	59	1	(1.69)	1976
1977	273	3	(1.10)	1977
1978	26	2	(7.69)	1978
1979	235	4	(1.70)	1979(2), 1980(2)
1980	296	2	(0.69)	1980
1981	124	0	(0)	
1984	277	2	(0.72)	1984(1), 1985(1)
1986	139	2	(1.44)	1987
1989	312	1		
Total	1,846	23	(1.25)	

Table 6. Long-distance recovery dates of Southern Giant Petrels (*Macronectes giganteus*) banded as nestlings near Palmer Station

	Number of birds recovered according to month								
	Apr.	May	June	July	Aug.	Sept.	Oct.	Nov.	Unknown
Year 1			2	10	3	1	1	1	1
Year 2	1		1			1			

Table 7. Recovery rates and sex ratios of Southern Giant Petrels (*Macronectes giganteus*) banded as nestlings and later recovered as breeding adults near Palmer Station

Date banded	No. banded	Recovered No.	(%)	Earliest recovery date	male	female	unsexed
1974–75	105	4	(3.81)	1980–81	1	3	0
1975–76	59	2	(3.39)	1983–84	1	0	1
1976–77	273	27	(9.89)	1983–84	10	9	8
1977–78	26	0	(0)				
1978–79	235	14	(5.96)	1985–86	6	6	2
1979–80	296	17	(5.74)	1984–85	6	9	2
1980–81	124	10	(8.06)	1985–86	6	4	0
1983–84	277	2	(0.72)	1988–89	2		
1985–86	139	0	(0)				
Total	1,534	76	(4.95)		32	31	13

far from its natal site on 30 December 1989 at its nest in the Joubin Islands by S. Poncet (personal communication). No. 1217-00540, banded 12 February 1980 by SJM and NPB on Cormorant Island, was recovered far from its natal site on 26 November 1988 at its nest in the South Shetland Islands at Harmony Point (62°19' S, 59°15' W), Nelson Island, by Marco Favero (personal communication). The meager returns outside the Palmer area are surprising in view of the extensive observations of giant petrels by others in the South Shetland Islands, indicating a high degree of philopatry for the Palmer population, and possibly a high mortality of young as well.

Most of the giant petrels at Palmer nested on Stepping Stones, Norsel Point, Shortcut, Hermit, and Humble islands, judging by the total number of young banded over the years (Table 8). However, the numbers of breeding pairs at those and other breeding sites varied considerably between seasons due primarily to absenteeism of certain adults, as indicated by Table 9. Although the number of breeding pairs fluctuated at specific sites, the total number of pairs remained fairly constant for the entire Palmer study area.

Breeding age. The Palmer birds of known age and sex ranged from four to twelve years old when found nesting for the first time. Although giant petrels were not

Southern Giant Petrel

checked during 1981-83, when some young almost certainly bred for the first time, the data show that both males (7.4 percent of the sample) and females (3.5 percent) bred as early as four years of age. However, 44 percent of the males were found breeding for the first time when seven years old, and 39.2 percent of the females when eight years old.

A nestling banded 578-01796 by others at Bonaparte Point on 18 February 1965 was first observed nesting in this study on 4 February 1975, when rebanded 658-67513 by DRN and WRF. When it was twenty-six years old it was observed with its chick at Bonaparte Point by DFP on 9 February 1991.

Satellite Tracking. In this study MRF initiated a satellite tracking program—evidently the first of its kind dealing with a seabird (Parmelee et al. 1985; Strikwerda et al. 1986). Bird-borne transmitters weighing about 200 grams were developed by the Johns Hopkins Applied Physics Laboratory and designed to work with the ARGOS system. Male giant petrels weighing more than 5 kilograms were good subjects not only because of their size, but also because they were easily obtained at their nests on Humble Island within the Palmer study area. During 19-27 January 1985, six males were harnessed with transmitters, and all of their

Figure 7.

Map of the Antarctic Peninsula showing the routes of two male Southern Giant Petrels (*Macronectes giganteus*) tracked by satellite during early 1985 (after Strikwerda et al. 1986).

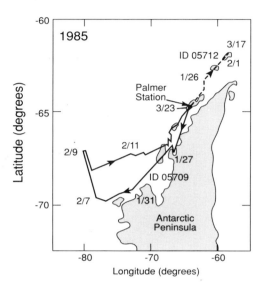

flights were tracked by satellite for about two months before signals from all six birds ceased (Table 10).

One male traveled more than 2,000 kilometers well out over the South Pacific, whereas the others flew up or down the Antarctic Peninsula. Two returned frequently to their nests (Figure 7). To date none of these birds has been recovered, and their fates and those of the transmitters they carry remain a mystery. The project demonstrated that seabirds can be tracked by satellite to obtain information otherwise unobtainable. Future experiments will have improved harness designs and substantially lighter transmitters with multiple sensors.

Summer Records. Since giant petrels are prone to follow ships at sea, they were observed in many places in Palmer Archipelago. Breeding pairs were, however, very restricted, being mostly confined to the southern coastal areas of Anvers Island. By far the largest concentration was within the Palmer study area, where approximately 300 pairs bred annually (Figure 8). It was difficult to assess the area's total population because for reasons unknown many adults from the marked population did not return to breed every year. Since the overall numbers of breeding pairs did not vary much between seasons, this kind of absenteeism would have been easily overlooked if not

Table 8. Total count of Southern Giant Petrels (*Macronectes giganteus*) banded as nestlings within the Palmer study area for 1976-77, 1978-79, 1979-80, 1983-84, and 1988-89

Locality	Range in no. of nestlings	Mean no. of nestlings
Stepping Stones	49–72	62.2
Norsel Point	46–75	60.8
Shortcut Island	49–61	34.6
Hermit Island	21–33	29.2
Humble Island	21–37	27.4
Cormorant Island	13–22	17.4
Limitrophe Island	5–16	12.6
Bonaparte Point	5–12	07.0
Laggard Island	4–08	05.6

Note: Numbers of pairs fluctuated considerably between seasons at specific sites due to periodic absenteeism.
[a] Based on nestlings two weeks old or older. Allowing for a 5–10% mortality of eggs and/or small chicks, the actual nest counts for these breeding sites would be somewhat higher.

Table 9. Philopatric history of 74 Southern Giant Petrels (*Macronectes giganteus*) banded as nestlings and later recovered as breeding adults at or close to their natal areas within the Palmer study area

Banding locality	No. banded	Total recovered No.	(%)	natal site M	F	U	close by M	F	U
Stepping Stones	452	17	(3.76)	5	6	3	2		1
Norsel Point	333	11	(3.30)	3	1	2	2	3	
Shortcut Island	235	11	(4.68)		1		4	5	1
Humble Island	227	9	(3.96)	1	2		4	2	
Hermit Island	159	4	(2.52)	1	1			1	1
Litchfield Island	129	7	(5.43)	2	1		2	2	
Cormorant Island	102	7	(6.86)	3				2	2
Limitrophe Island	74	2	(2.70)					2	
Bonaparte Point	68	6	(8.82)	1	1		2	1	1
Total	1,779	74	(4.16)	16	13	5	16	18	6

Note: M = male; F = female; U = sex uncertain.

for the banded individuals. Site tenacity was strong, and although quite a few birds skipped nesting some years, they later returned to breed at their traditional sites.

An uncertain number of pairs were distributed unevenly elsewhere along the southern coasts of Anvers Island. DFP noted a few nesting individuals in the Joubin Islands on 16 and 23 January 1975, and again on 12 January 1984; a few kilometers north of Cape Monaco he recorded 27 nestlings and one egg on 5 February 1979. S. Poncet (personal communication) also noted nesting birds in these same areas, including 17 nests in the Joubin Islands on 30 December 1990. Only a scattering of nonbreeding individuals and flocks were noted at all other land areas visited in Palmer Archipelago.

Winter Records. BMG banded and color coded adults at nests on Bonaparte Point and Humble Island and continued to observe them during the winter of 1977 and into their next breeding season. Although the species was present throughout his residency, its numbers fluctuated as individuals left the area and returned at various times. Marked individuals invariably returned to their breeding sites and did not fly from island to island as did some unmarked, presumably nonbreeders. Num-

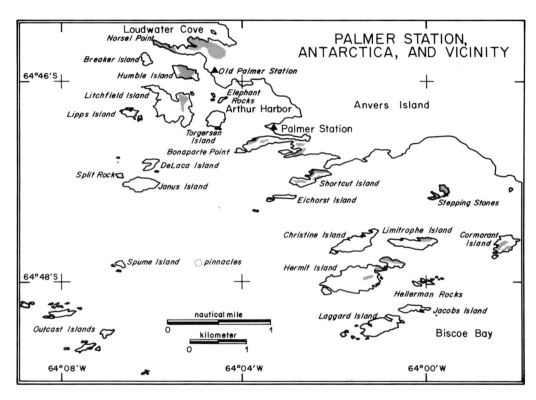

Figure 8. Major breeding areas (approximate locations indicated by areas shaded with parallel lines) of Southern Giant Petrels (*Macronectes giganteus*) within the Palmer study area. Minor and often sporadic breeding areas were Laggard, Jacobs, Janus, DeLaca, and Breaker islands, and Hellerman and Elephant rocks.

Table 10. Time periods and distances traveled by six male Southern Giant Petrels (*Macronectes giganteus*) from nests located near Palmer Station (from Strikwerda et al. 1986)

Trans. No.	Dates	Distance (km)
05701	26 Jan. 85–02 Feb. 85	188
05703	19 Jan. 85–19 Mar. 85	1,082
05705	28 Jan. 85–08 Mar. 85	1,150
05709	21 Jan. 85–17 Mar. 85	2,320
05712	21 Jan. 85–17 Mar. 85	639
05714	21 Jan. 85–16 Mar. 85	1,730

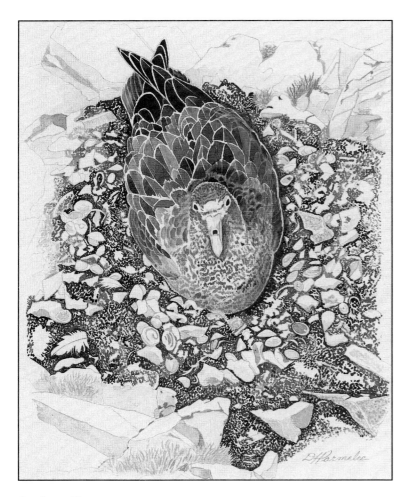

**Southern Giant
Petrel**

bers of giant petrels diminished following the fledging of young in May, and their numbers continued to drop until mid-August, when there was a fresh insurgence of birds that extended into the breeding season. Similar lows and highs were noted during 1955-57 by Holdgate (1963), also in 1976 by WRF, and in 1979 by NPB.

Many unbanded giant petrels visited BMG's Bonaparte Point colony from mid-August in 1977. The birds wandered about, nest scraped at sites not used by breeders the previous season, and elicited agonistic behavior from marked birds at nests. BMG concluded that it was the failed breeders and prebreeders that visited the breeding grounds in winter to prospect for sites and mates. One banded failed breeder from the previous season wandered about the colony and displayed before one of the unbanded birds.

Heimark and Heimark (1984) counted 102 giant petrels on Shortcut and Stepping Stones islands on 3 July 1983 and estimated that 80 percent of these birds were paired at traditional nests; they continued to see pairs at nests throughout the winter. During WinCruise I, Pietz and Strong (1986) recorded the species between 22 August and 22 September 1985 along transects extending from 56°30' N to 66°30' S, including several runs through Palmer Archipelago. Most sightings were of one to three birds that followed the ship; the largest concentration observed was seven birds at 64°28' S, 61°44' W, in Gerlache Strait on 25 August. Both adults and subadults were noted at sea along the transects, also at the breeding grounds on Humble Island on 30 August.

Color Phases. Southern Giant Petrels showed a wide range in plumage color from nearly black to nearly white, gray being predominant. Not uncommon were gray birds with pale, sometimes nearly white, heads. An earlier estimate by Parmelee et al. (1977b) on the proportion of white-phase birds was much too high, and BMG's estimate of 2 percent for the Palmer area is more accurate. BMG also noted that only 2 percent of the chicks produced in 1977 were white-phased.

Irides of the giant petrels also showed a wide range, from very dark brown to bright yellow, including much flecking in the many observed in this study. All but one of the white-phased birds had very dark irides; the exception had yellow, somewhat flecked irides.

Feeding Behavior. Observations on the feeding behavior of Southern Giant Petrels were limited. Not only were the birds indifferent to seal carcasses, but, unlike the skuas and gulls, they were rarely seen at the open dumps before the latter were closed at Palmer Station. One concludes that all the giant petrels left the breeding areas to forage mostly at sea, where they were observed feeding on whale carcasses, krill, garbage—almost anything available to them. However, Hunter (1983) found significant dietary differences between the sexes at South Georgia, where females fed more extensively at sea on krill, squid, and fish, while the males fed more on carrion.

Clues to the summer foods taken by the giant petrels at Palmer derived mostly from items regurgitated by nestlings. In 1977, BMG found fish, bird, and numerous seal skin and fur remains in the regurgitations, indicating that the wide-ranging adults were finding carcasses outside of the breeding areas. In 1985, CCR concluded that the chicks were fed krill before mid-February, but thereafter penguin remains predominated until his departure on 7 March. In 1989, WRF found numerous remains of penguin chicks in the regurgitations of young giant petrels after the young penguins had left their breeding grounds that year. Since giant petrels seldom visited the penguin rookeries, this led to speculation as to where they preyed on the penguins. Following the oil spill at

White-phased Southern Giant Petrel at Litchfield Island nest site. This was the only white individual noted with pale irides; all others were dark. Photographed 15 January 1975.

Arthur Harbor in late January 1989, thousands of young Adélie Penguins were contaminated upon swimming out to sea from the nesting areas and, according to WRF, may have provided the giant petrels with a windfall of carcasses.

With respect to winter birds, Pietz and Strong (1986) examined the stomach contents of four specimens they collected during winter in the seas south of Anvers Island. At 65°21' S, 65°03' W, on 2 September 1985, an adult male contained only feather parts, while an immature female contained two squid beaks and feather parts. At 65°08' S, 65°48' W, on 9 September 1985, an adult male contained three squid beaks, feather parts, and seal fat and hair, while an adult female contained three squid beaks and feather parts. All four specimens were only moderately fat (1.1 on a scale of 0 to 4).

Predatory Behavior. Little predatory behavior by giant petrels was observed at the Palmer breeding grounds. During the

Incubating Southern Giant Petrel nearly inundated with fresh snow. Although giant petrel eggs survived the spring blizzard, those of Antarctic Terns nearby were abandoned. Photographed on Bonaparte Point, 23 November 1975.

1974-75 season, several pairs of giant petrels nested close to nesting Antarctic Terns without apparent show of anxiety by either species. One tern nest succeeded within a mere 5 meters of a giant petrel nest. Only on rare occasions did the terns pursue flying or grounded giant petrels, though they flew at the skuas relentlessly. Such behavior by the terns indicated that nesting giant petrels posed no threat. Crippled individuals apparently triggered predatory responses; SJM watched a giant petrel snatch an injured adult tern from the surface of the sea and devour it on the spot.

Mortality. As a rule, Southern Giant Petrels were tenacious in defense of their eggs and small chicks. Males and females alike were easily hand caught at the nest, and this facilitated banding and periodic checking. The species' main defenses were its strong, hooked bill, capable of inflicting painful wounds, and its ability to fling noxious regurgitated foods and oils for a meter or more—a trait that has made the giant petrel famous. Among the many vigilant defenders were a few timid individuals that were prone to fly off in the presence of humans and leave their eggs or small chicks unattended. The unattended often disappeared, presumably taken by skuas, perhaps by other giant petrels, though this was not documented. Holdgate (1963) noted predation of giant petrel eggs by skuas at Arthur Harbor on 28 November 1958, but gave no details.

Egg mortality due to human intrusion probably was less than 5 percent at the beginning of incubation, and it decreased as the season advanced. For reasons unknown, more flushes than usual occurred during snowstorms. During a spring blizzard of 20-21 November 1975, nine out of twenty-nine birds left their eggs on one of the islands of Stepping Stones—truly an extraordinary number of flushes. On the other hand, the large majority of incubating birds at Palmer remained on their eggs, despite the drifting snow that covered all but their heads. The eggs of all these close sitters survived.

Nestlings were zealously protected by the parents for the first two weeks, and noticeably less thereafter. Three-week-old chicks were left unattended at a time when they seemed to be immune from predatory attack. Like the adults, the chicks flung copious amounts of oil at human intruders; this appeared to be their main defense against avian predators as well. According to WRF, an adult Kelp Gull that had strolled too close to a giant petrel chick was suddenly doused in oil; the gull then spent considerable time and energy in an attempt to clean its plumage. On rare occasions adults viciously defended large chicks; one parent literally charged DFP with open bill and wings, forcing a hasty retreat.

The banding, weighing, and measuring of adults resulted in no mortalities, but one female evidently died from heat exhaustion when MRF and DFP took too long in placing a harness on its body for telemetry purposes. Many of these birds began to pant after ten minutes of handling, and once back on the nest they continued to pant for some time. No birds were lost during twenty minutes of handling, but that was the maximum time we allowed ourselves. Handling resulted in fear responses and increased heart rate and temperature. Downy chicks often panted on unusually warm days. According to physiologist

D. Murrish (personal communication), the birds were heat stressed.

In February 1979, Arthur Harbor endured a severe fowl cholera epidemic that was particularly injurious to its small population of Brown Skuas, several specimens of which were autopsied. The finding of several giant petrel carcasses (none autopsied) in the area at that time suggested that the species also suffered to some extent from the disease. Other than this singular experience, no adult giant petrel carcasses were found within or outside of the Palmer study area.

Breeding Biology. The breeding biology of Southern Giant Petrels has been studied elsewhere by several ornithologists, notably Warham (1962), Voisin (1968, 1978), Conroy (1972), Mougin (1968, 1975), and Hunter (1984b). In this study BMG's observations paralleled theirs, with minor differences and additions. Some behaviors not witnessed by BMG were attributed to an apparent indifference to feeding, and consequent displaying, at seal carcasses by his birds. Many of the species' behaviors associated with breeding were observed throughout the winter period; they were intensified from mid-August, when large numbers of pairs and individuals were at the nesting sites.

Nest sites. Ranged from a few to over 50 meters above sea level, mostly on high, prominent rocks exposed to the wind. North-facing slopes were favored over the south-facing ones that had perpetual snow.

Density. Isolated pairs were not uncommon; in some cases there was only one pair per island, but most pairs nested colonially. The densest concentrations occurred on Stepping Stones, where as many as 41 nests have been scattered more or less uniformly over one of its two islets. Some nests on Bonaparte Point were only 3.5 meters apart.

Site tenacity. According to Mougin (1968) and Conroy (1972), giant petrels tend to return to the same nest year after year. Marked adults in this study did not return to the Palmer breeding area every year, but upon returning they usually occupied the same island or peninsular nesting ground, not necessarily the same nest. BMG noted that during the winter of 1977 a pair built a new nest about 46 meters from one used by them the previous season. Another pair abandoned the nest used by them the previous season and simply displaced a defending male from its previously occupied nest nearby; the displaced bird disappeared after several failed attempts to establish another nest.

Changing one island site for another was exceptional: two males nested on Bonaparte Point before moving a short distance to Shortcut Island, where they acquired new mates. Another exceptional case was enigmatic. An incubating female checked by WRF at Litchfield Island on 24 December 1988 was found two days later quite some distance away on a nest at Shortcut Island. Conroy (1972) mentions giant petrels occupying nests belonging to others of their kind, but behavior of this sort appears to be unusual, and it is little understood.

Site defense. Agonistic behavior observed by BMG included low-intensity threats by birds that waved their heads from side to side from a sitting position, accompanied by vocalizations. At times the heads were thrust out horizontally along the ground, as though poised to strike a blow with the bill—evidently the "lunging forward" display described earlier by Warham (1962). According to BMG, the lunges were directed at birds merely passing close to the defended site, not necessarily at those approaching it directly.

Not all the species' ritualized displays were noted in this study, particularly the "forward-threat" described by Warham (1962) and usually observed at disputed seal carcasses. The "upright-threat" was observed frequently by BMG and others of this study whenever a flying individual landed at a defended site, or when ground

Agonistic "upright-threat" display.

intruders approached the site. The displaying bird stood with wings extended and neck stretched vertically, accompanied by vigorous head waving and braying.

Encounters rarely resulted in wounds. However, BMG noted two marked males from neighboring nests locked in vigorous and bloody combat at Humble Island on 31 October 1977—some days before any eggs appeared.

Sexual displays. BMG rarely observed aerial displays in early winter (24 May, 9 and 17 June, 16 July 1977 recorded), but commonly in late winter from 19 August. The displaying bird stretched its neck and brayed while in flight, sometimes repeatedly, usually above others on the ground, but at times when no others were in sight. One performing male suddenly descended and engaged its mate in mutual display. According to BMG, the aerial display elicited agonistic responses as well as sexual ones.

Ground displays usually consisted of the one described by Warham (1962) as the "greeting," in which one or both members of a pair waved the head and brayed while approaching one another. The greeting was often followed by "mutual displaying" that occurred when the two indulged in varying amounts of bill touching, nibbling, mutual preening, and bowing; according to BMG, the display was often repeated, generally every one to four minutes at a sitting. Although WRF observed mutually displaying pairs in 1976 from May through August, he noted them more frequently in October and November. BMG observed that in 1977 mutual displaying was uncommon following fledging from mid-May to early August, but common from mid-August until shortly before egg-laying in early November. NPB noted mutual displaying in September and October 1979.

Sexual displaying long before egg-laying probably reinforced the pair-bond and enhanced the opportunity for early breeding by experienced adults. Breeding success in some seabirds has been shown to be greater in more experienced pairs (Carrick and Ingham 1970), and Conroy (1972) noted that established pairs of giant petrels tended to lay earlier than those breeding for the first time. Presumably, young from early nestings have a better chance of escaping winter food shortages. Cooperation by a strongly bonded pair probably results in better chick survival as well.

Conroy (1972) observed that males tended to arrive at the nest sites before females, and occupied their nests more often than the females in August and September. BMG noted that with rare exceptions single birds at nest sites in winter were males.

Nest-scraping. BMG noted that both sexes cleared the nest of accumulated snow by scraping it away with their bills from either a sitting or standing position, often tossing chunks to one side. At times nest-scraping took place during mutual displaying. Holdgate (1963) noted nest-scraping activity at Arthur Harbor as early as late July in 1955; the birds disappeared following a fresh snowfall on 18 August and were not observed again until 29 August. In 1977, BMG observed little nest-scraping for a long period following blizzards and increased snow depth. His records indicated that ice cover and low temperatures had little to do with giant petrel abundance, but that snow cover influenced nest-scraping and other activities related to breeding. Conceivably, the birds conserved energy by delaying nest-scraping until after the worst of the winter snows.

Nest-building. Southern Giant Petrel nests were often large, conspicuous mounds composed mostly of limpet shells (deposited earlier by gulls) and pebbles, interspersed with pieces of lichens, chunks of moss, and bits of grass; a few were built almost entirely of vegetative material. According to BMG, both sexes participated and built from either a sitting or standing posture. Sitting individuals simply dropped material over their shoulders, while those standing tossed material backwards by vigorous turns of the head. Unlike nest-scraping activities that occurred early, the actual moving of limpet shells and other materials was not observed before late October, within a fairly short time of egg-laying.

Copulation. Copulation has been observed infrequently. Holdgate (1963) noted copulation at Arthur Harbor in mid-October 1957. BMG noted it there on 17 September and 24 October in 1977 and observed that the male climbed onto the sitting female and stropped his bill vigorously across that of its mate. Intruding males not only initiated mutual display behavior but attempted to rape incubating females, according to BMG, who observed that females often successfully defended themselves, but at times succumbed to the advances.

At point Géologie, Adélie Land, Bretagnolle (1989a) observed that copulation in giant petrels is not introduced by a special display, nor does it complete a courtship. Males will attempt copulation at any time. Successful copulation, however, usually takes place following diminution of aggressivity and a synchronization and increasing similarity of male and female courtship—at a time when females replace appeasement posture with head-shaking displacement activity.

Egg-laying. The observed range in egg-laying was 2-26 November, but most eggs were laid toward the middle of the month (Parmelee et al. 1977b).

Clutch size. One.

Incubation period. From laying to hatching, the period taken for several eggs ranged from 59 to 64 days, with a mean of 61 days (WRF). According to Hunter (1984b), the range for 90 eggs on South Georgia was 57-66 days, with a mean incubation period of 60 days.

Incubation. Both sexes took part, but it was not determined how much time each incubated. Presumably the unbanded birds that occupied eggless nests for long periods were prebreeders, but also the failed breeders that had lost eggs continued to sit on their empty nests.

Brooding. Both sexes participated in brooding, but it was not determined how much each brooded. Chicks were brooded and guarded closely for two weeks. After the third week they were left completely unattended for long spells. Within five days of hatching, giant petrel chicks acquired their homeothermy and were capable of enduring temperatures between 0° and 10°C (Ricklefs 1982). The long brooding period possibly may function as insurance against severe summer storms; almost certainly it is a predator deterrent.

Hatching. The range in hatching dates was 4-30 January. Most young hatched toward the middle of the month (11-16 January), reflecting the mid-November peak in egg-laying.

Prefledging. Nestlings left the nest several days before they could fly. Holdgate (1963) observed that prefledged chicks waddled down to the sea near Norsel Point as late as 16 May in 1955. The following year he observed that chicks, presumably fledged, departed from Norsel Point during 3-10 May.

Fledging. The range in fledging dates was 1-19 May (DRN, BMG). Although the fledging period from hatching to strong flying was not determined from marked individuals in this study, it likely approximated 117 days in some cases, since hatching occurred as early as 4 January. According to

Hunter (1984b), the mean fledging period was 123 days for 15 males and 117 days for 18 females on South Georgia.

Synopsis of annual cycle. Although adult Southern Giant Petrels were common year-round residents, in Palmer Archipelago they bred principally along the south coast of Anvers Island. Most young of the year migrated to far-distant wintering grounds, but many returned to their natal areas to breed when four years of age or older, both sexes exhibiting a high degree of philopatry. Sexual activities took place on the breeding sites in winter, though they temporarily halted during periods of heavy snow. Site and mate tenacity were generally high. Individuals did not always return to breed annually, and such absenteeism resulted in fluctuations in the number of breeding pairs from season to season at specific sites. Most pairs bred colonially. The range in egg-laying was 2-26 November. The incubation period ranged from 59 to 64 (mean 61) days; the fledging period was an estimated 117 days. The range in fledging dates was 1-19 May. The period from first egg to last fledging was 198 days—by far the longest recorded for any species in Palmer Archipelago.

Southern Fulmar
(*Fulmarus glacialoides*)

Overall Breeding Distribution. Circumpolar. The Antarctic and Maritime Antarctic including the main continent, Antarctic Peninsula, and Peter, South Shetland, South Orkney, South Sandwich, and Bouvet islands.

Current Status in Palmer Archipelago. Common year-round resident with widely scattered colonies. The existence of unknown colonies seems probable. Known colonies need additional assessments.

Summer Records. Southern Fulmars were rare in the vicinity of Palmer Station in summer, but throughout this study they were encountered in Dallmann Bay and Gerlache Strait, where their numbers indicated large colonies nearby. Parmelee and Rimmer (1985) suspected breeding after sighting birds flying about the high cliffs of northern Brabant Island on 30 December 1983, and their nesting was later confirmed by S. Poncet (personal communication), who noted many landing on ledges on both sides of the strait between Brabant and Davis islands on 16 March 1986. Poncet reported finding additional colonies in northern Palmer Archipelago, including "large" one at Cape Wollaston (63°40' S, 60°47' W), Trinity Island, and one of uncertain size not far away at Tower Island on 22 January 1987. She also found a small one that year on the northwest coast of Anvers Island (location and date obscure). Much earlier, Holdgate (1963) reported on Southern Fulmar sightings at sea off Anvers Island's northwest coast in December 1956.

Winter Records. Small and large numbers of Southern Fulmars were seen at various times near Palmer Station during early winter, and far less often during late winter. Sightings were negligible for 1955-57, for Holdgate (1963) recorded but one observation—a single bird on 20 March 1956. DRN noted thousands flying over Bismarck Strait in 1975 on 8 May; thereafter their numbers decreased greatly from 9 to 18 May, and none was seen again until 28 September. WRF noted small numbers in 1976 on 5 April, hundreds by 12 April, and thousands by 18 April, when they peaked; thereafter their numbers dropped off steadily until 21 May, when they were last recorded that year. NPB noted a flock near Bonaparte Point on 8 June 1979.

During WinCruise I in 1985, Pietz and Strong (1986) observed Southern Fulmars from 56°30' S to 65°30' S. Within Palmer Archipelago, they recorded the following: in Bransfield-Gerlache straits (between 63°32'

S and 64°16' S), twelve singles and a group of three during 27-28 August; at Dallmann Bay (between 64°14' S and 64°13' S), seven singles on 16 September (unpublished notes).

Diet. Obst (1985) observed that although Southern Fulmars fed regularly on krill and were among the best avian indicators of krill presence, the importance of krill in their overall diet was not known. The species also fed on other kinds of prey, for a

specimen obtained in Gerlache Strait by DFP contained the remains of fish on 22 October 1975.

Breeding Biology.

Nest sites. Sites observed at Palmer Archipelago were inaccessible ledges on vertical cliffs overlooking the sea. Those not far away on Aspland Island (61°30' S, 55°49' W) in the South Shetlands were similarly situated, although some of the nesting ledges there were accessible without climbing equipment. The single eggs at that colony were laid on bare rock.

Southern Fulmars feeding in the nearly wide-open waters of Gerlache Strait. Photographed 1 December 1975.

Incubating Southern Fulmar in its nesting niche high on a sea cliff. Photographed on Aspland Island in the South Shetlands on 30 December 1977.

Egg dates. Six eggs examined by DFP at the Aspland Island colony on 30 December 1977 each contained a very small embryo. Not all pairs observed closely had eggs, indicating that laying had only recently begun that season.

Incubation. Both sexes incubated. Incubating individuals and accompanying mates were reluctant to leave their eggs when approached closely and, once flushed, they were quick to return. Oil ejection appeared to be their defense.

Antarctic Petrel
(*Thalassoica antarctica*)

Overall Breeding Distribution. Confined to the Antarctic on or close to the main continent. Possibly breeds on the Balleny Islands.

Current Status in Palmer Archipelago. Transient; rare in summer, uncommon to abundant at other times. No evidence of nesting within or near Palmer Archipelago.

Summer Records. Small numbers of migrants were observed near Palmer Station during early spring. DRN recorded one to four birds in 1975 on 2, 4, 9, and 28 October, followed by fifteen scattered individuals among many Snow Petrels that flew westward low over the glacier behind the station on 30 October. A male collected on 30 October had somewhat enlarged gonads. NPB recorded one near the station on 1 October 1979. Elsewhere in Palmer Archipelago, DFP noted one at the leading edge of the pack ice at Dallmann Bay on 27 October 1975.

Midsummer records were lacking and late summer ones nearly so. DRN and WRF noted one over Arthur Harbor near Litchfield Island on 27 February 1975. DFP noted singles over Bransfield Strait at 63°37' S, 61°10' W, and at 63°51' S, 61°25' W, on 1 March 1990.

Winter Records. Antarctic Petrels have been recorded near Palmer Station from April to October, but most frequently during late winter, when variable numbers of

Table 11. Winter sightings of Antarctic Petrels (*Thalassoica antarctica*) at Palmer Archipelago during WinCruise I, 22 August–22 September 1985 (Pietz and Strong, unpublished notes)

No. of individuals	Date	General location	Coordinates
19	25 Aug. 85	Gerlache Strait	64°21′ S, 61°50′ W, to 64°32′ S, 62°31′ W
20	27 Aug. 85	Bransfield Strait	63°30′ S, 61°32′ W, to 63°55′ S, 61°11′ W
40	28 Aug. 85	Bransfield/Gerlache straits	63°25′ S, 60°58′ W, to 64°16′ S, 61°35′ W
1	14 Sept. 85	Bismarck Strait	64°44′ S, 64°23′ W
29	16 Sept. 85	Dallmann Bay	64°14′ S, 62°38′ W, to 64°13′ S, 62°41′ W
13	17 Sept. 85	Dallmann Bay/ Bransfield Strait	64°10′S, 62°42′ W, to 64°03′ S, 62°44′ W
41	18 Sept. 85	Bransfield Strait	63°33′ S, 62°13′ W, to 63°11′ S, 61°08′ W

migrants appeared. Holdgate (1963) recorded small numbers during 14-29 September 1957. DRN noted the following in 1975: fifty-one birds on 26 April, twenty-two on 18 May, two on 5 July, three on 30 July; literally "thousands" during 15-21 September, with reduced numbers by 22 September, and only one remaining by 23 September. During early winter of 1976, WRF noted one to five birds occasionally, but none after 20 July. NPB noted the following in 1979: one on 8 August, two on 9 August, one on 10 August, and one on 1 October. During the winter of 1983, Heimark and Heimark (1984) noted a "few" between 17 July and 7 September. Only in 1975 did anyone witness mass migrations, but conceivably such migrations may have occurred beyond view of shore-based personnel.

During WinCruise I, Pietz and Strong (1986) recorded Antarctic Petrels between 57°30' S and 67°30' S during 22 August-22 September 1985. Unpublished notes on their sightings at Palmer Archipelago during that cruise are summarized in Table 11.

Diet. During WinCruise I, Pietz and Strong (1986) collected thirteen male and one female Antarctic Petrels in the seas south of Anvers Island between 64°55' S and 66°11' S latitudes during 6-15 September 1985. Alimentary tracts from these specimens contained the following: five birds had the remains of fish, six had krill, three had unidentified crustaceans, nine had squid, one had amphipods, and one had pteropods. The one female lacked only squid and amphipods, indicating a similar diet to that of males. According to PJP, some of the krill were identified as *Euphausia superba,* some of the fish as *Pleuragramma.*

Body Weight. PJP obtained the weights and determined the fat contents for fourteen Antarctic Petrel specimens collected during WinCruise I. Weights of thirteen males ranged from 560 to 750 grams (average, 664.4); for one female, 667. On a fat scale of 0 to 4, nine males ranged from 0 to 2, averaging 1.0 (light); the one female was 3 (heavy).

Cape Petrel
(*Daption capense*)

Overall Breeding Distribution. Circumpolar. Breeds across all antarctic and subantarctic life zones from the main continent to islands off New Zealand.

Current Status in Palmer Archipelago. Year-round resident; uncommon breeder, abundant winter migrant. Recorded every month near Palmer Station. Although Cape Petrels were dispersed widely in summer throughout Palmer Archipelago, only a few breeding sites were found. The birds were often conspicuous at sea, probably owing to the species' propensity to circle and follow ships. Enormous numbers of winter migrants occurred over Bismarck Strait.

Flocking Cape Petrels in the oily waters by the carcass of a Southern Bottle-nosed Whale. Photographed in Bransfield Strait on 4 January 1978.

Summary Records. Cape Petrels bred abundantly in the South Shetland Islands, but their numbers fell off sharply in Palmer Archipelago, where apparently nesting had not been documented before this study. The behavior of several adults near Anvers Island's Quinton Point at 64°20' S, 63°38' W, on 3 February 1979 led DFP, SJM, and NPB to suspect nesting. On 27 December 1983, nesting on Anvers Island was confirmed by DFP and CCR with the finding of an incubating bird near Giard Point at 64°25' S, 63°53' W. At Brabant Island that same season, Parmelee and Rimmer (1985) recorded 35 Cape Petrels nesting at about 64°08' S, 62°36' W, near Astrolabe Needle

on 30 December; they also noted hundreds of birds that probably bred between the Needle and Metchnikoff Point. At Trinity Island, DFP recorded 6 incubating Cape Petrels near Skottsberg Point (63°55' S, 60°49' W) on 14 January 1984; that same day birds were noted but nesting was not confirmed at Farewell Rock (63°52' S, 61°01' W). S. Poncet (personal communication) observed other possible breeding areas on Trinity Island, and at Tower Island as well. Concentrations of Cape Petrels should be looked for along the west coast of Liège Island, for DFP, MRF, and B. Obst recorded three incubating birds at a sea cliff in a bay south of Moureaux Point (63°57' S, 61°49' W) on 31 January 1985.

Cape Petrels were scarce in areas visited along the Antarctic Peninsula. DFP and CCR recorded four incubating birds at Spigot Peak (64°38' S, 62°34' W) on 31 December 1983, and DFP observed two nest ledges with a chick each near the Argentine Almirante Brown Station, Paradise Bay, during the 1988-89 and 1989-90 seasons.

Winter Records. Overwintering observers at Old Palmer and U.S. Palmer Station recorded Cape Petrels on numerous occasions. Holdgate (1963) stated that during 1955-57, individuals and small flocks of Cape Petrels were seen periodically near Arthur Harbor in spring, autumn, and winter. With the aid of a high-powered telescope mounted at Bonaparte Point, DRN recorded Cape Petrels throughout the winter of 1975. He observed 189 birds during 9-24 April, and thousands flying westward over Bismarck Strait during 26 April-11 May. Thereafter, he recorded smaller numbers, including 185 birds during 12-20 May, 26 during 27-29 May, 61 during 1-8 June, 93 during 21-27 June, 1 on 22 July, none during August, and 3 during 25-29 September. During the winter of 1976, WRF noted small groups of Cape Petrels over Bismarck Strait on 19 April, followed

Table 12. Winter sightings of Cape Petrels (*Daption capense*) at Palmer Archipelago during WinCruise I, 22 August–22 September 1985 (Pietz and Strong, unpublished notes)

No. of individuals	Date	General location	Coordinates
2	25 Aug. 85	Gerlache Strait	64°28′ S, 61°52′ W, to 64°30′ S, 62°12′ W
1	26 Aug. 85	Anvers Island, Fournier Bay	64°33′ S, 62°49′ W
2	27 Aug. 85	Orleans Strait	63°57′ S, 60°41′ W
10	27 Aug. 85	Orleans/Gerlache straits	64°00′ S, 60°55′ W
5	27 Aug. 85	Gerlache Strait	63°55′ S, 61°11′ W
8	27 Aug. 85	Bransfield Strait	63°45′ S, 61°20′ W, to 63°30′ S, 61°32′ W
14	28 Aug. 85	Bransfield Strait	63°32′ S, 60°52′ W, to 63°54′ S, 61°18′ W
3	28 Aug. 85	Gerlache Strait	64°06′ S, 61°22′ W
5	28 Aug. 85	Gerlache Strait	64°16′ S, 61°35′ W
2	31 Aug. 85	Bismarck Strait	64°49′ S, 64°00′ W
1	14 Sept. 85	Bismarck Strait	Dream Island
5	14 Sept. 85	Bismarck Strait	64°44′ S, 64°23′ W
3	15 Sept. 85	Bismarck Strait	64°49′ S, 64°04′ W
42	16 Sept. 85	Dallmann Bay	64°20′ S, 62°42′ W, to 64°13′ S, 62°41′ W
10	17 Sept. 85	Dallmann Bay	64°10′ S, 62°42′ W
8	17 Sept. 85	Bransfield Strait	64°03′ S, 62°44′ W

Incubating Cape Petrel in a crevice on a low sea cliff. Photographed on Nelson Island in the South Shetlands on 20 November 1973.

by thousands on the 23 April, when their numbers peaked; numbers declined rapidly thereafter, until by 20 July only a few individuals and small groups remained. During the winter of 1979, NPB noted only 53 Cape Petrels during all of April, but then recorded "thousands" over Bismarck Strait on 2 May; thereafter, only 4 during the remainder of May, only 8 throughout June, and no more until 15 October. During the winter of 1983, Heimark and Heimark (1984) saw only a "few" Cape Petrels during 17 April–22 May, a few during 1-16 July, and infrequently after 4 September. Although Watson et al. (1971) did not include the area south of Anvers Island in

their map folio distribution for the species, Bismarck Strait appears to be an important migration route for Cape Petrels, and possibly an important feeding area as well (Parmelee et al. 1977b).

On WinCruise I, Pietz and Strong (1986) recorded the species between 56°30′ S and 66°00′ S during 22 August-22 September 1985. Table 12 summarizes their unpublished notes for sightings pertaining to Palmer Archipelago.

Diet. Specimens obtained by Pietz and Strong (1986) during WinCruise I indicated that Cape Petrels fed on a variety of prey during the winter of 1985. Of three males collected at 65°00′ S, 63°22′ W, in Flandres

Bay south of Wiencke Island on 29 August, the first contained one euphasid and two squid beaks, the second eight euphasids and one squid, and the third nine euphasids and five squid beaks. A fourth specimen collected at 64°55′ S, 64°24′ W, about 25 kilometers south of Anvers Island on 15 September, contained nine fish otoliths.

Predation. The remains of a Cape Petrel's egg were found beside a Brown Skua's nest at the petrel colony near Astrolabe Needle, Brabant Island, on 29 December 1983 (Parmelee and Rimmer 1985). In the absence of penguins, the Brown Skuas at this site had to feed on the Cape Petrels and possibly other seabirds.

Breeding Biology. The relatively few nest sites recorded in Palmer Archipelago were essentially like the many examined in this study at Nelson, Deception, and Aspland islands in the South Shetlands.

Nest sites. Single-egg clutches were laid in unlined crevices or on narrow ledges at various heights on the sides of sea or inland cliffs. Approaches to some sites were easily negotiable; others were so inaccessible that special equipment would have been necessary for close observation. Incubating birds were reluctant to leave their eggs when approached closely; their main defense was ejecting oil. According to A. Mazzotta (personal communication), a Cape Petrel was recovered in January 1991 at its nest site where banded 28 years previously in 1965 at Deception Island, South Shetland Islands.

Breeding chronology. One egg examined closely at Giard Point, Anvers Island, was fresh on 27 December 1983. That same season egg-laying had just commenced at the colony near Astrolabe Needle, Brabant Island, on 29 December. In the South Shetlands, at Harmony Cove, Nelson Island, fresh eggs were noted as early as 20 November 1973, and, judging by the number of pairs at eggless sites, egg-laying had not yet peaked.

Snow Petrel
(*Pagodroma nivea*)

Overall Breeding Distribution. Circumpolar. Breeds at numerous localities on or near the main continent and Antarctic Peninsula, but also on South Orkney, South Sandwich, South Georgia, Bouvet, Balleny, and Scott islands.

Current Status in Palmer Archipelago. Year-round resident; uncommon breeder in summer, uncommon to abundant at other times. Observed mostly at sea in the presence of pack ice.

Summer Records. Although Holdgate (1963) failed to record Snow Petrels near Arthur Harbor in summer, occasional sightings of individuals and even small flocks were recorded there throughout summer by observers in this study, but the species' breeding areas remained obscure. According to Holdgate, the sighting of two Snow Petrels flying above Mount Francois on 1 December 1956 suggested that small numbers of them bred high up in the eastern ranges of Anvers Island, but this was never confirmed. A more plausible breeding area appeared to be the rugged, island-dotted northwest coast of the island, since literally

Table 13. Winter sightings of Snow Petrels (*Pagodroma nivea*) at Palmer Archipelago during WinCruise I, 25 August–18 September 1985 (Pietz and Strong, unpublished notes)

No. of individuals	Date	General location	Coordinates
45	25 Aug. 85	Gerlache Strait	64°21′ S, 61°50′ W, to 64°32′ S, 62°31′ W
Many	26 Aug. 85	Fournier Bay, Anvers Island	64°33′ S, 62°49′ W
87	27 Aug. 85	Gerlache/ Bransfield straits	63°57′ S, 60°41′ W, to 63°30′ S, 61°32′ W
96	28 Aug. 85	Gerlache/ Bransfield straits	63°25′ S, 60°58′ W, to 64°16′ S, 61°35′ W
Small nos.	30 Aug. 85	Bismarck Strait	near Palmer Station
2	14 Sept. 85	Bismarck Strait	64°44′ S, 64°23′ W
24	15 Sept. 85	Bismarck Strait	64°49′ S, 64°04′ W, to 64°55′ S, 64°24′ W
78	16 Sept. 85	Dallmann Bay	64°29′ S, 62°49′ W, to 64°13′ S, 62°41′ W
31	17 Sept. 85	Dallmann Bay	64°10′ S, 62°42′ W, to 64°03′ S, 62°44′ W
28	18 Sept. 85	Bransfield Strait	63°33′ S, 62°13′ W, to 63°11′ S, 61°08′ W

thousands of Snow Petrels migrated westward low above the glacier behind Palmer Station during 30-31 October in 1975. Specimens taken from the steady stream that flew there had much enlarged gonads. Likewise, Heimark and Heimark (1984) recorded "thousands" flying westward behind Palmer Station during 28-29 October 1983, and it seemed likely that the birds were headed for some undescribed breeding ground. Several cruises along the island's western and northern coasts disclosed no breeding colonies large or small, thus heightening the mystery surrounding these migrating hordes, since only open sea lay westward from there.

Probable breeding colonies for Palmer Archipelago occurred along the northern coast of Brabant Island. In 1983, Parmelee and Rimmer (1985) observed a Snow Petrel enter a presumed nest crevice among nesting Cape Petrels at a cliff near Astrolabe Needle on 29 December; the following day they recorded 10 Snow Petrels at presumed nest crevices high on cliffs near Duclaux Point (64°04' S, 62°15' W), and not far from there upward of 40 of them at presumed nest crevices on towering cliffs at both sides of a narrow strait between Brabant and Davis islands. On 31 December Parmelee and Rimmer observed 6 Snow Petrels at

Snow Petrel

Pair of Snow Petrels in their nesting crevice on a sea cliff. Photographed on Signy Island in the South Orkneys on 18 December 1974.

there in 1975, when peak numbers occurred on 24 April (hundreds) and several times during late May and June (thousands). In 1976, WRF noted the species on 12 April (hundreds) and on 18 April (thousands). NPB recorded it during every month throughout the winter of 1979: 1 to 2 individuals on 19, 22, and 24 April; 2 to 6 individuals on 21, 25, and 31 May, and 2, 3, and 27 June; 1 to 12 individuals on 13, 20, 21, 22, and 31 July, also 3, 4, 8, and 10 August, and 2, 7, and 9 September; "many all day" on 10 September, and 2 to 6 individuals on 12, 16, and 17 September. In 1983, Heimark and Heimark (1984) observed a flock of 30 to 40 birds near Palmer Station on 2 May, and from that time the species was observed at least every week throughout winter; peak numbers occurred on 12 June.

During WinCruise I, Pietz and Strong (1986) recorded Snow Petrels between 62° S and 67° 31' S during 25 August to 18 September 1985. Unpublished notes on their sightings at Palmer Archipelago during that cruise are summarized in Table 13.

Diet. According to Zink (1981a), the preferred foraging habitat of Snow Petrels, as with the Adélie Penguin, is an ice concentration of 3-5 oktas (1-4 oktas = light

crevices high on the cliffs of Spigot Point, Antarctic Peninsula, about 8 nautical miles from Brabant Island.

Winter Records. Holdgate (1963) stated that during 1955-57 there were fairly frequent records of Snow Petrels at Arthur Harbor between 27 March and 5 November. DRN noted that their numbers fluctuated

pack ice, 5-8 = heavy pack ice), especially pack ice several years old, with irregular, chambered surfaces that provide shelter for resting and molting birds and a matrix-type habitat for krill and other prey species. The Snow Petrels follow these ice edges closely, exhibiting high maneuverability with a peculiar short but rapid wing beat described by Watson (1975) as "batlike." Zink believed that this kind of foraging behavior was consistent with the observation of Falla (1964) that the Snow Petrel's primary food consists of dead or injured macroplankton that accumulates at the edge of ice.

No doubt macroplankton is an important source of food, but fish appear to be equally important, at least during late winter. During WinCruise I, Pietz and Strong (1986) collected 32 Snow Petrels for food analysis between 26 August and 15 September 1985. Except for 2 males and a female taken at Fournier Bay, Anvers Island, on 26 August, all remaining specimens were taken from seas further south, between 64° 58' S and 66°11' S latitudes. Of 29 birds of known sex, males outnumbered females 25 to 4. Alimentary tracts from all 32 specimens contained the following foods: 19 birds had the remains of fish, 18 had krill, 2 had unidentified crustaceans, 1 had squid, and 1 had tissue believed to have been scavenged from a dead seal. According to PJP, some of the krill were identified as *Euphausia superba*. Fish and krill appeared to be the two most important foods for both sexes.

Fraser et al. (1989) observed Snow Petrels that were attracted to krill swarms near the ice edge that extended from South Island to Deception Island in the South Shetlands during 18-19 June 1987. Unlike the Antarctic Terns, the petrels foraged chiefly over ice, but they also flew over open water near the ice edge while taking advantage of krill swarms. Of the seven species of birds recorded near the ice edge, only Snow Petrels were observed foraging at night—still another advantage, since the krill moved closer (vertically) to the surface during the dark hours (1600-1800).

Body Weight. PJP obtained the weights and determined the fat contents for 29 Snow Petrel specimens collected during WinCruise I. Weights of 25 males ranged from 210 to 375 grams (average, 297.4); 4 females ranged from 270 to 310 grams (average, 266.5). On a fat scale of 0 to 4, 22 males ranged from 0 (trace) to 3 (heavy), averaging 1.9 (medium); 4 females ranged from 1 (light) to 2, averaging 1.25.

Blue Petrel
(*Halobaena caerulea*)

Overall Breeding Distribution. Transitional and cold Sub-Antarctic, including South Georgia, Prince Edward, Marion, Crozet, and Kerguelen islands; also on offshore stacks near Macquarie Island, and possibly on Diego Ramirez in the Drake Passage.

Current Status in Palmer Archipelago. Rare pelagic visitor in summer and winter; possibly a much overlooked species, especially in northern Palmer Archipelago.

Summer Records. DFP, MRF, and B. Obst observed one at sea being harassed by a South Polar Skua at about 63°58' S, 61°52' W, off the northeast coast of Liège Island on 31 January 1985.

Winter Records. During WinCruise I in 1985, Pietz and Strong (unpublished notes) recorded the species' latitudinal distribution between 56°30' S and 63°42' S. On 28 August they observed one at 63°42' S, 61°07' W, west of Trinity Island; on 18 September one at 63°33' S, 62°13' W, northwest of Hoseason Island, and two at 63°18' S, 61°26' W, northeast of Hoseason Island (unpublished notes).

Family of Storm-Petrels

Of the five species of breeding storm-petrels listed for the Antarctic and Sub-Antarctic by Watson (1975), only the Wilson's Storm-Petrel (*Oceanites oceanicus*) was an abundant summer resident in Palmer Archipelago. The Black-bellied Storm-Petrel (*Fregetta tropica*) nested commonly not far away in the South Shetland Islands, but it was observed very infrequently in the archipelago, where it is not known to breed.

Although Wilson's Storm-Petrel was not targeted as a prime species to be considered in this study, Roberts's (1940) early classic on its breeding biology was produced in the Argentine Islands only 40 nautical miles southeast of Palmer Station. Beck and Brown (1972) later made additional contributions at Signy Island in the South Orkneys. While based at Palmer Station, Murrish and Tirrell (1981) conducted physiological experiments on the species, and Obst (1985, 1986) and Obst et al. (1987) focused on its foraging behaviors and food preferences.

Incubating Wilson's Storm-Petrel. Several loose stones were temporarily removed to reveal the secluded nest chamber within rock rubble. Photographed at Bonaparte Point, Anvers Island, 1 January 1984.

Wilson's Storm-Petrel
(*Oceanites oceanicus*)

Overall Breeding Distribution. Circumpolar. Breeds in the Antarctic and Maritime Antarctic, and in transitional, cold, and temperate (possibly) subzones of the Sub-Antarctic. Subspecies *O. o. oceanicus* breeds in South Georgia, Cozet, Kerguelen, Falkland, and Tierra del Fuego islands, and possibly also on Balleny, Peter, and Bouvet islands. *O. o. exasperatus* breeds from South Sandwich Islands to numerous localities on the Antarctic Peninsula and main continent.

Current Status in Palmer Archipelago. Wilson's Storm-Petrel was the most abundant and widely dispersed procellariid recorded in summer throughout Palmer Archipelago, where at the southern sector it arrived in November and departed by early May. Migration dates for the northern sector were not determined.

Banding. A total of 162 Wilson's Storm-Petrels were banded within the Palmer study area, and an additional 3 not far away on Dream Island in Wylie Bay.

Adults. Of 145 adults banded, 44 were captured at nest sites: during 1974-75, at Bonaparte Point (14 birds); during 1977-78, at Bonaparte Point (11); during 1983-84, at Bonaparte Point (6), Cormorant Island (4), Dream Island (3), and 1 each at Laggard, James, DeLaca, Litchfield, and Humble islands and Norsel Point. The remaining 101 adults were captured in mist nets: during 1983-84, at Litchfield Island (64), Hermit Island (22), and Bonaparte Point (15). There have been no long-distance recoveries to date. Between-season recaptures at the same nest site indicated site fidelity, but these studies were not intensified because of the high incidence of nest abandonment by incubating birds following handling and banding.

Nestlings. Of 20 nestlings banded, 4 were banded at Litchfield Island during 22-24 January 1978; 8 at Bonaparte Point and 4 at Humble Island on 10 March 1980; 1 each at Bonaparte and Norsel points and Litchfield and Humble islands during 20 February-7 March 1984. There have been no recoveries to date.

Summer Records. Sightings of Wilson's Storm-Petrels at innumerable suitable breeding habitats indicated that the birds were widespread and abundant throughout Palmer Archipelago, as they were in the South Shetlands and at most places visited along the Antarctic Peninsula. Although numerous nests were found within the Palmer study area, far fewer were recorded elsewhere, because there was only limited time to search for them. At Brabant Island Parmelee and Rimmer (1985) recorded two nests with eggs near Astrolabe Needle, and another with an egg at Metchnikoff Point on 30 December 1983. DFP observed three nests with eggs on Dream Island on 5 January 1985.

Arrival. One individual observed by DFP at sea near Smith Island (63°00' S, 62°30' W) as early as 21 October in 1975 indicated that Wilson's Storm-Petrels approach Palmer Archipelago long before they reach southern Anvers Island, where their arrivals have been documented. Holdgate's (1963) early arrival dates ranged from 16 to 25 November for 1955-57. A few were at Bonaparte Point in 1973 on 22 November, and scores had arrived by 28 November (Parmelee et al. 1975). WRF first observed one at Bonaparte Point in 1975 on 13 November, and scores by 15 November. In 1979, G-AM noted one at Bonaparte Point on 14 November, and many by 18 November. Heimark and Heimark (1984) first observed one near Palmer Station in 1983 on 22 November.

Following the arrivals, individuals were observed vocalizing on rocks in the open. Evidently these were males that were advertising for mates by means of "chattering calls" described in detail by Bretagnolle (1989b), who studied the species elsewhere.

Departure. Holdgate's (1963) latest departure dates for the Palmer area ranged from 16 April to 8 May for 1955-57. In 1975, DRN last observed a nestling at Bonaparte Point 9 April, although he observed storm-petrel tracks in the snow at another site as late as 27 April. WRF last recorded storm-petrels of any age at Bonaparte Point on 23 April in 1976. Palmer Station personnel recorded the species flying near the station on 24 April 1979, and Heimark and Heimark (1984) last recorded it there on 14 April in 1983.

Foraging Behavior. In comparing the foraging behaviors of several seabird species in the Palmer area, Obst (1985) stated that foraging trips by Wilson's Storm-Petrels lasted from two to four days, thus substantiating earlier observations by Beck and Brown (1972). By multiplying the storm-petrel's traveling speed of 48.4 kilometers per hour by the mean foraging trip duration of 60.8 hours, and dividing the total in half, he estimated that the potential foraging range was 744 kilometers—the distance achieved if all of the time were spent flying from and back to the nest along straight paths. Although the distance is exaggerated, it exemplifies the species' great mobility to search out and exploit available patches of krill. Even a single euphausiid represents substantial energy for the small Wilson's Storm-Petrel. According to Obst, it would take 24 minutes for a storm-petrel to use the energy assimilable in 1 gram of krill (two to four adult euphausiids), but less than 3 minutes for an albatross to use the same quantity.

Surface-feeding by the storm-petrels has been widely observed (e.g., Zink and Eldridge 1980), but underwater feeding has been recorded much less often. While diving near Palmer Station, W. Hamner (personal communication) observed that Wilson's Storm-Petrels submerged completely (just beneath the surface) to grasp rising oil droplets before the droplets could float to the surface and burst.

Diet. Obst (1985) stated that although Wilson's Storm-Petrel is an opportunistic feeder and will take a wide variety of food when available, it feeds heavily upon krill (*Euphausia superba*) during the summer months. The proportion of krill by weight in its diet amounted to as much as 85 percent. By using captive birds, Obst (1986) demonstrated that the species has the ability to digest wax. Inasmuch as waxes are abundant constituents of many marine organisms, he concluded that the ability to digest wax was an important adaptation for the Wilson's Storm-Petrel and other marine birds that encounter waxy prey, allowing these birds to benefit from a vast source of potential energy.

Mortality. The remains of a small number of adult Wilson's Storm-Petrels were noted at South Polar and Brown skua sites. A few found in the passageways between cliffs on Litchfield Island suggested that the skuas somehow caught the birds as they flew along the narrow corridors. Pietz (1987) noted that one pair of South Polar Skuas was very adept at catching adult storm-petrels that nested in the skuas' territory on Bonaparte Point. Little evidence was found that the skuas penetrated storm-petrel nest chambers, but one situated in moss had been picked apart and the egg eaten. Skuas migrated from the region about the time the young storm-petrels fledged, but whether they preyed on emerging fledglings during that period was not determined. Presumably those storm-petrels that fledged late escaped skua predation.

Entombment by spring and fall snowstorms appeared to have been the most important mortality factor—amply documented by the mummified remains of adults, eggs, and young in the nest chambers. Beck and Brown (1972) concluded that exposed habitats subjected to wind scour had much better storm-petrel nesting productivity relative to those that were sheltered. In recognizing that storm-petrels face entombment at both ends of their breeding season, Beck and Brown believed that a portion of the young would survive each year because of the spread in egg-laying. They reasoned that in some years birds laying early would lose their eggs through snow blockage, whereas chicks from later layings would probably survive if there were few autumn snows, and vice versa. Observations in this study suggest that the majority of the Palmer population delayed egg-laying and favored a midcycle peak in egg-laying, with early and late nestings being exceptional. These hypotheses merit further study.

Breeding Biology.

Nest sites. Wilson's Storm-Petrels required small crevices for their egg chambers deep enough to be beyond the reach of avian predators, notably skuas. These were situated in glacial rubble, screes, and deep fissures, and also beneath boulders in tunnels presumably excavated by the birds in soft ground, moss, guano, and volcanic cinders. The nest sites occurred from near the edge of the sea to the tops of sea cliffs and inland ridges. Although usually well hidden from view, their presence and approximate locations were easily detected through the birds' vociferous vocalizations. These are probably best identified as "grating" calls. Bretagnolle (1989b) described variants of these calls that differed in temporal and frequency characteristics for each sex, indicating that they played a role in sexual recognition. Males also had a long version of the grating call that was used in agonistic interactions.

Not many nest chambers could be reached without considerable rock removal and other disturbances. Since the storm-petrels readily occupied artificial sites consisting of loose piles of stones, including old cairns, several were constructed on Bonaparte Point to facilitate nest observations.

Nest Chambers. The single egg was sometimes laid on bare ground, but more often on materials either deposited by wind and water or dropped by the birds. It was not determined whether the birds intentionally hauled nesting material. However they were deposited, the materials included pebbles, bits of mosses and lichens, limpet shells, and shed penguin feathers, as well as old storm-petrel feathers, egg shells and membranes, and, not infrequently, mummified young.

Density. Although many observers agreed that Wilson's Storm-Petrel was an abundant breeder in the vicinity of Palmer Station, estimates ranged widely concerning their total numbers for any given season. Most estimates were based on the number of adults seen fluttering low over the nesting grounds during the twilight hours early in the season; others were based on the number of nest sites located. Scores of these birds could be seen at once flying about Bonaparte Point, but a total count of the breeding population was not determined even for this one area because of the time and effort required to locate breeding pairs. A more accurate measure could have been obtained by the use of mist nests, as demonstrated by CCR, who in less than three hours caught sixty-two storm-petrels in a net at Litchfield Island as late as 5 March in 1984. Beck and Brown (1972) concluded that although it was not possible to access a total population by this method, it did provide a relative measure of the number of birds in flight over an area between seasons.

Within the Palmer study area nearly all islands and peninsulas had breeding pairs, but most nesting sites were recorded at Bonaparte and Norsel points and at Litchfield, Cormorant, and Hermit islands, where there was much suitable habitat. Isolated nestings were not uncommon, but often several pairs nested only a few meters apart in prime situations.

Egg-laying. Although large numbers of storm-petrels occupied their nest chambers soon following the arrival of the early vanguards, several weeks passed before the first eggs appeared. One plausible explanation is that the delay in egg-laying ensured against entombment by late snowstorms. Beck and Brown (1972) also believed that sufficient time is required to produce the species' large egg.

The onset of egg-laying varied somewhat between seasons at marked Bonaparte Point nests; for example, it started nearly a week earlier in 1983-84 than in 1984-85. At no time were eggs recorded before 24 December, though a chick that hatched there on 20 January in 1975 indicated egg-laying as early as 12 December that season. Judging by the several hundred eggs observed over the years, early laying was exceptional. Most eggs were laid between 24 December and 10 January, and not many thereafter. A fresh egg was recorded at Bonaparte on 14 January 1978, and another was laid there sometime after 16 January in 1979 (DFP).

Incubation. The banding of adults at nest sites in the study area showed that both sexes incubated, but the period of incubation was not determined. Elsewhere, Roberts (1940) found that it varied from thirty-nine to as many as forty-eight days. Beck and Brown (1972) later discovered that attentiveness by the incubating birds accounted for the variation.

Hatching. DFP noted a newly hatched chick on Bonaparte Point in 1975 as early as 20 January, while PJP noted one on Humble Island in 1980 as late as 24 February. Allowing a minimum of thirty-nine days for incubation, the egg that was laid no earlier than 16 January in 1979 likely hatched no sooner than 24 February.

Fledging. WRF recorded the fledging period at one Bonaparte Point site: a chick that hatched on 26 January 1976 left the nest chamber sixty-seven days later, on 2 April. This period is near the upper end of a series of twelve fledgings ranging from fifty-four to sixty-nine days (Beck and Brown 1972). Allowing an average sixty days for fledging, some Bonaparte Point chicks probably fledged as early as 19 March, and others as late as 25 April. It seems likely that the majority fledged in early April, given the rather narrow midseason peak in egg-laying.

Synopsis of annual cycle. Wilson's Storm-Petrels arrived at their breeding grounds along the south coast of Anvers Island from mid- to late November. The vanguards, usually few in number, were soon followed by many that flew low over the nesting areas during twilight hours. The nesting habitat was exposed rocky areas from near sea level to high inland ridges. Egg-laying was delayed for several weeks, during which time pairs vocalized loudly within their well-hidden nest chambers in crevices or burrows beyond the reach of predators.

Pairs used the same nest chamber from one season to the next, depositing their single egg on bare ground or on accumulated debris. Egg-laying ranged from mid-December to mid-January, but most eggs were laid between 24 December and 10 January. Entombment by spring and autumn snows was the greatest threat to the survival of storm-petrels at their nest sites, and possibly accounted for the delay in laying. Both sexes incubated. Foraging trips lasted 2 to 4 days, krill being the principal food, and wax also being utilized. Hatching occurred from 20 January to about 24 February. The one recorded fledging period of 69 days falls near the upper limits of those recorded for the species elsewhere. Fledging occurred from about 19 March to 24 April, although most young probably fledged in early April. The period from first egg to last fledging was about 134 days. Few storm-petrels of any age were present beyond mid-April, but one time they were recorded as late as 8 May.

Black-bellied Storm-Petrel
(Fregetta tropica)

Overall Breeding Distribution. Circumpolar. Breeds in the Maritime Antarctic and transitional, cold, and temperate (possibly) subzones of the Sub-Antarctic. Subspecies *F. t. tropica* breeds on South Shetland, South Orkney, South Georgia, Crozet, Kerguelen, and Antipodes islands, possibly also on Bouvet and Prince Edward islands. A white-bellied subspecies, *F. t. melanoleuca*, possibly breeds on Gough Island. The relationship between *F. tropica* and *F. grallaria* (called White-bellied Storm-Petrel) needs clarification.

Current Status in Palmer Archipelago. Rare pelagic visitor; no evidence of nesting. This is possibly an overlooked species, particularly in the northern sector of Palmer Archipelago.

Summer Records. WRF observed one at sea near the Joubin Islands on 1 February 1975. On 19 March that year DRN banded one that had been captured by hand on the Palmer Station site. Palmer's closest known nesting ground was Deception Island in the South Shetlands, where DFP located an incubating female with a freshly laid egg on 3 January 1978. Conceivably these storm-petrels forage commonly in Bransfield Strait, for DFP counted as many as 19 there between 62°19' S, 58°33' W, and 62°25' S, 58°38' W, on 27 January 1990. All these birds flew in the direction of Palmer Archipelago, possibly having come from breeding areas on the South Shetland Islands.

Family of Cormorants

*S*hag is another name for cormorant. Although the names shag and cormorant are commonly applied to certain species or even local populations, the two names are often used interchangeably. The family consists of about thirty extant species. It is widely distributed in the world, being absent only on remote oceanic islands and in polar regions that remain mostly frozen year-round. Relatively few family members have established breeding colonies in the Southern Ocean. The more northern King Shags (*Phalacrocorax albiventer*) are resident in the Sub-Antarctic on Marion, Crozet, Kerguelen, and Macquarie islands; they are also found in Chile, Argentina, and the Falkland Islands. Blue-eyed Shags (*P. atriceps*) overlap with King Shags in South America, but otherwise breed separately on Shag Rocks, South Georgia, South Sandwich, South Orkney, South Shetland, and Heard islands, as well as in the vicinity of the Antarctic Peninsula, including Palmer Archipelago.

Morphologically, King and Blue-eyed shags are very similar in size and appearance and are separable in the field notably by minor differences in white cheek patterns and the presence or absence of white dorsal patches and wing bars. Some taxonomists consider the two as morphs of a single species: "King Shag morphs" and "Blue-eyed Shag morphs." Lumped together they are called "Imperial Shags." Whether one or two species, various populations are reduced further to subspecies levels separable from one another by minor characteristics, such as bill size. The population that occupies Palmer Archipelago is distinguished by having long tails and short bills, hence the scientific trinomial *Phalacrocorax atriceps bransfieldensis*—the Antarctic Blue-eyed Shag.

Complicating the nomenclature further is a recent publication by Siegel-Causey and Lefevre (1989), who, on the basis of five diagnostic osteological features, propose that the Antarctic Blue-eyed Shag represents a

Antarctic Blue-eyed Shags on Cormorant Island.

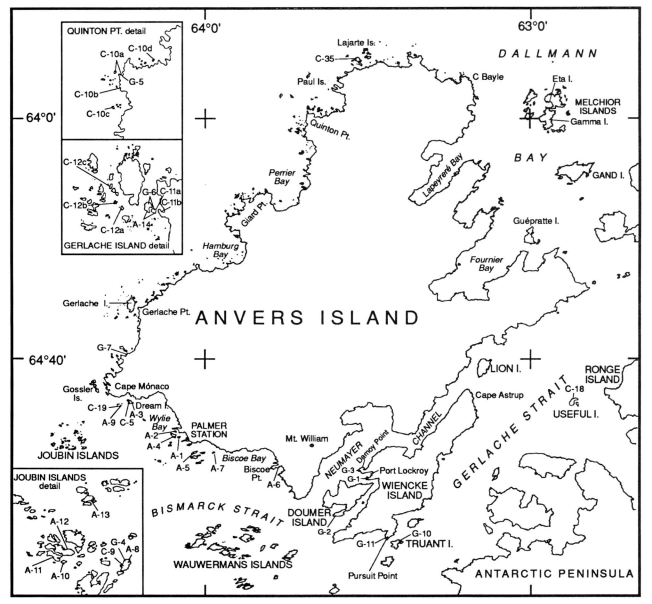

Figure 9. Approximate colony locations of Antarctic Blue-eyed Shags (*Phalacrocorax atriceps bransfieldensis*) in Palmer Archipelago.

separate genus: *Notocarbo bransfieldensis*. Until this taxonomic matter is settled, I prefer to use the trinomial designation; the common name, Antarctic Blue-eyed Shag, sets it apart from all other populations. If readers find it difficult to manage the taxonomic jargon, they will rejoice in knowing that the Antarctic Blue-eyed Shag can hardly be confused with any other shag or bird species that inhabits Palmer Archipelago.

Antarctic Blue-eyed Shag
(*Phalacrocorax atriceps bransfieldensis*)

Overall Breeding Distribution. Clarification is needed on the relationships of the shags in the sub-antarctic and antarctic regions. *P. a. bransfieldensis* breeds chiefly, if not exclusively, in the Maritime Antarctic.

Current Status in Palmer Archipelago. Antarctic Blue-eyed Shags were abundant, conspicuous, and one of the more uniformly distributed breeders throughout the South Shetland Islands, Palmer Archipelago, and along the Antarctic Peninsula to about 68° S latitude. The birds occurred mainly in sight of land and were not prone to follow ships. At least part of the population that inhabited Palmer Archipelago was resident throughout the year, and pairs occupied their traditional breeding sites in winter. Whether part of the population migrated from the region during winter is not known, but local movements to open water during hard freezes were apparent. In this study, most of the observations on shags were carried out during two seasons at a colony near Palmer Station by Maxson and Bernstein (1980, 1982), Bernstein and Maxson (1981, 1982a, b, 1984, 1985), and Bernstein (1982).

Banding. A total of 589 Antarctic Blue-eyed Shags were banded within the Palmer study area at Cormorant Island.

Adults. During 1978-81, SJM and NPB banded 225 adults, including many that were sexed (92 males, 78 females) and color coded for time-budget studies. To date there have been no recoveries of any of these banded birds beyond Cormorant Island.

Nestlings. During 1978-81, SJM and NPB banded 206 chicks, including 3 that were caught following fledging. In 1983-84, CCR banded an additional 158 chicks. To date there have been no recoveries of any of these birds beyond Cormorant Island and Christine Island nearby, despite many searches for them at other colonies near and far.

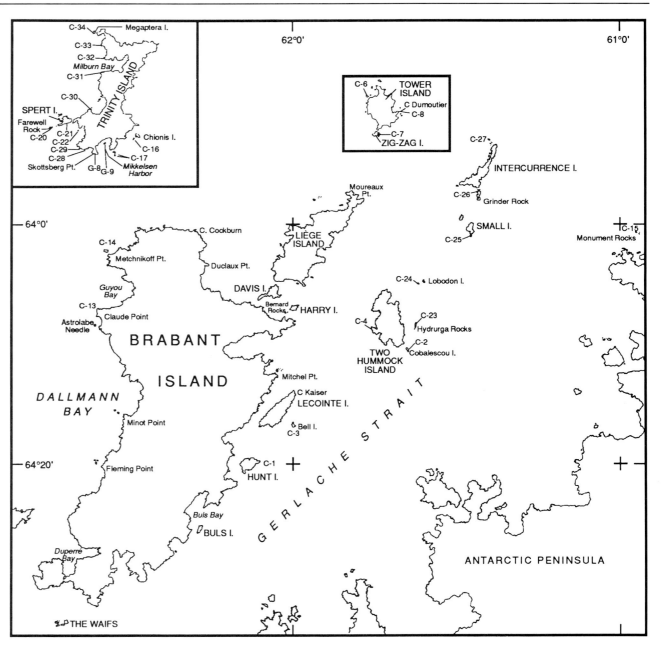

Table 14. Approximate colony locations of Antarctic Blue-eyed Shags (*Phalacrocorax atriceps bransfieldensis*)

No.	Location	Count	Nature	Date	Reference
S-1	Cormorant Island	100	N_3	25 Nov. 73	DFP & SJM
	(64°48′ S, 63°58′ W)	7	N_1	28 Dec. 74	DFP
		285	N_1	09 Jan. 78	DFP
		485	N_1	31 Dec. 78	SJM & NPB
		326	N_1	19 Dec. 79	SJM & NPB
		378	N_1	08 Dec. 83	H & H (1984)
		729	N_1	02 Dec. 85	H & H (1988)
S-2	Cape Astrup	30	P_3	09 Nov. 74	DFP
	(64°43′ S, 63°11′ W)	50	N_3	21 Jan. 83	S. Poncet
S-3	Bay Point	15	P_3	10 Nov. 74	DFP
	(64°46′ S, 63°20′ W)	28	N_1	20 Jan. 83	S. Poncet
S-4[a]	Joubin Islands	68	N_1	16 Jan. 75	DFP & F. Todd
	(64°45′ S, 64°23′ W)	325	C_1	05 Feb. 79	DFP, SJM & NPB
S-5[a]	W of Quinton Pt	17	C_1	02 Feb. 79	DFP, SJM & NPB
	(64°21′ S, 63°42′ W)	11	N_1	04 Jan. 90	S. Poncet
S-6[a]	W of Quinton Pt	25	C_5	02 Feb. 79	DFP, SJM & NPB
	(64°22′ S, 63°42′ W)	45	$P_{3/4}$	04 Jan. 90	S. Poncet
S-7[a]	W of Gerlache Pt	79	N_1	03 Feb. 79	DFP, SJM & NPB
	(64°36′ S, 64°15′ W)	124	N_1	02 Jan. 85	DFP & JMP
S-8	Port Lockroy	60	N_3	19 Jan. 83	S. Poncet
	(64°50′ S, 63°31′ W)	65	N_1	28 Dec. 83	DFP & CCR
		40	N_3	26 Jan. 84	S. Poncet
		60	N_3	05 Jan. 89	DFP
S-9	Priest Island	43	N_1	22 Jan. 83	S. Poncet
	(64°52′ S, 63°31′ W)	54	N_1	29 Dec. 84	DFP & JMP
S-10[a]	W of Brabant I	6	N_5	29 Dec. 83	DFP & CCR
	(64°19′ S, 62°34′ W)				
S-11[a]	W of Brabant I	6	N_5	29 Dec. 83	DFP & CCR
	(64°14′ S, 62°36′ W)				
S-12	Claude Point	24	N_1	29 Dec. 83	DFP & CCR
	(64°06′ S, 62°37′ W)				
S-13[a]	SE of Brabant I	23	N_1	30 Dec. 83	DFP & CCR
	(64°29′ S, 62°35′ W)				
S-14[a]	E of Brabant I	7	N_1	30 Dec. 83	DFP & CCR
	(64°27′ S, 62°23′ W)				
S-15[a]	E of Brabant I	22	N_1	30 Dec. 83	DFP & CCR
	(64°25′ S, 62°18′ W)				
S-16	Lecointe Island	52	N_3	30 Dec. 83	DFP & CCR
	(64°13′ S, 62°03′ W)				
S-17[a]	E of Brabant I	2+	N_5	30 Dec. 83	DFP & CCR
	(64°13′ S, 62°03′ W)				

Some of these birds banded as nestlings returned to the natal colony as prebreeders and later as breeding adults. Although not checked during 1981-83, the following were recorded during 1983-85. One-year-olds simply visited the colony. Some three- and four-year-olds paired and defended unlined, eggless nests. In three known cases, four-year-olds raised young. In nine cases, five-year-olds raised young, while several others of the same age appeared not to be nesting.

Shaw's (1981, 1985a, b) large sample of banded shags from the South Orkney Islands included breeding individuals ranging in age from three to twelve years. Difficult to explain was his observation that 29 percent of his birds selected mates of the same age—twice the number expected if mate selection occurred at random with respect to age. He concluded that the relationship between age and date of return is weak in the shags, and he also found no evidence of age-clumping within the colony. He hypothesized that partners select on the basis of their age—an interesting idea that merits further study.

Summer Records. In the later stages of this study an attempt was made to locate and map extant shag colonies throughout Palmer Archipelago. This proved only partially successful because some of the col-

in Palmer Archipelago.

No.	Location	Count	Nature	Date	Reference
S-18	Bernard Rocks (64°08′ S, 62°01′ W)	22	N_1	30 Dec. 83	DFP & CCR
S-19	E Skottsberg Pt (63°55′ S, 60°49′ W)	100+	N_5	14 Jan. 84	DFP
S-20	W Skottsberg Pt (63°55′ S, 60°50′ W)	100+	N_5	14 Jan. 84	DFP
S-21	Farewell Rock (63°51′ S, 60°55′ W)	30	N_5	14 Jan. 84	DFP
S-22	Pursuit Point (64°54′ S, 63°27′ W)	50	N_3	17 Jan. 84	S. Poncet
		140	N_3	06 Feb. 87	S. Poncet
S-23[a]	E of Wiencke I (64°49′ S, 63°15′ W)	20	N_3	17 Jan. 84	S. Poncet
S-24	Useful Island (64°43′ S, 62°52′ W)	12	N_3	17 Jan. 84	S. Poncet
S-25	Gamma Island (64°20′ S, 63°00′ W)	20	N_5	09 Dec. 84	DFP
		60	$P_{3/4}$	28 Jan. 87	S. Poncet
S-26[a]	E of Eta Island (64°19′ S, 62°53′ W)	40+	N_5	09 Dec. 84	DFP
		60	$P_{3/4}$	29 Jan. 87	S. Poncet
S-27[a]	E of Eta Island (64°19′ S, 62°53′ W)	5+	N_5	09 Dec. 84	DFP
		10	$P_{3/4}$	29 Jan. 87	S. Poncet
S-28[a]	W of Gerlache I (64°35′ S, 64°17′ W)	30	N_5	02 Jan. 85	DFP & JMP
		20	$P_{3/4}$	08 Feb. 87	S. Poncet
S-29	Christine Island (64°48′ S, 64°01′ W)	4	N_1	16 Jan. 85	DFP & JMP
		8	N_1	02 Dec. 85	H & H (1988)
		0	0	01 Jan. 89	WRF
		0	0	10 Jan. 90	DFP
S-30	Bell Island (64°16′ S, 61°59′ W)	200	N_3	22 Jan. 85	B. Obst (pers. comm.)
S-31[a]	E of Guepratte I 64°30′ S, 63°00′ W)	220	$P_{3/4}$	29 Jan. 87	S. Poncet
S-32[a]	Joubin Islands (64°45′ S, 64°23′ W)	250	P_4	09 Feb. 87	S. Poncet
S-33	Elephant Rocks (64°46′ S, 64°05′ W)	6	N_1	06 Jan. 90	WRF
S-34[a]	Lajarte Islands (64°15′ S, 63°24′ W)	30	$N_{4/5}$	05 Jan. 90	S. Poncet
S-35[a]	Lajarte Islands (64°15′ S, 63°22′ W)	20	$N_{3/4}$	05 Jan. 90	S. Poncet

Note: Colonies are numbered chronologically. N = nest; A = adult; P = pair; N_1 = individual nest counts accurate to ± 5%; N_3 = nest counts accurate to ± 10–15%; N_4 = nest counts accurate only to 50%; N_5 = guesstimate; H & H = Heimark and Heimark 1984, 1988; S. Poncet, personal communication.
[a] Nameless island, islet, or rock.

onies were seen from a distance, and their recorded coordinates and shag counts were approximations at best. Compounding the problem were the sharp fluctuations in numbers of breeding pairs between seasons, as was soon apparent in the Palmer study area at Cormorant Island, where numbers ranged from as few as 7 pairs to as many as 729. Some of the smaller colonies may have been incipient ones that conceivably disappear in time, as apparently was the case at the Christine Island colony in the study area. Extensive surveys throughout Palmer Archipelago by S. Poncet and J. Poncet added information through additional counts and more accurate coordinates. Much verification is still needed, and undescribed colonies remain to be discovered, especially in the northern sector of the archipelago. So far as is known, no assessments of accumulated guano deposits have been made at the Cormorant Island colony or others in the archipelago that show evidence of long usage—a study that might well reveal useful information on past occupancy.

In the case of shags, no attempt was made to include the often unpublished observations of early explorers. No doubt the colony at Port Lockroy and others on or near Wiencke Island were noted by them, as long ago as 1908 by French naturalists

(Gain 1914). Some of the early records are vague; for instance, Holdgate (1963) mentioned the colony at Cormorant Island, but evidently did not estimate the number of breeding pairs in 1955-57.

An updated, distributional summary combining the observations by members of this study and others, notably the Poncets (personal communication), is presented in Figure 9 and Table 14. Colony S-10 needs to be verified, since Parmelee and Rimmer (1985) reported on two colonies rather than one for that part of Brabant Island. Other colonies needing special attention are along the south coast of Trinity Island; in addition to S-19, S-20, and S-21, DFP noted several others that were deleted from Table 14 and Figure 9 because of insufficient information. Follow-up observations are needed for possible incipient colonies noted near Anvers Island by S. Poncet (personal communication): one pair at an empty nest on 19 January 1984 on Dream Island, where the species was not known to nest previously, and two pairs that possibly were nesting at a nameless islet south of Gerlache Island (between S-7 and S-22) on 8 February 1987.

Cormorant Island Colony. In this study there were roughly a hundred pairs at Cormorant Island when it was first visited on 25 November 1973. The main colony was on the upper levels of a sea cliff above two smaller subcolonies on sea cliffs nearby, and dispersed among the latter were small numbers of nesting Adélie Penguins. This arrangement did not vary much throughout the study, although the number of pairs fluctuated considerably between seasons.

When next checked on 28 December 1974, there was but a single occupied nest in the main colony and only six occupied nests at one of the subcolonies. Of the seven nests, four held fairly large young, and three had eggs only. No additional nestings occurred at Cormorant Island that season. Numbers of breeding pairs had swelled by 9 January 1978: 285 pairs at occupied nests. Bernstein and Maxson (1984) recorded 485 pairs on 31 December 1978 and 326 on 19 December 1979. Heimark and Heimark (1984) recorded 378 pairs on 8 December 1983; DFP, 587 pairs on 13 December 1984; and Heimark and Heimark (1988), 729 pairs on 2 December 1985.

Christine Island Colony. Parmelee et al. (1985) reported on an incipient shag colony that had been established during the 1984-85 season on Christine Island, only 2 kilometers from Cormorant Island. Thirteen individuals of uncertain age and seventeen yearlings (one banded) were present on 16 and 18 January. Several of the birds that sat on freshly constructed nests laid no eggs that season. Heimark and Heimark (1988) visited the colony next on 2 December 1985, finding eight nests with eggs that later produced thirteen young. No banded birds were reported. WRF visited the colony during 1988-89, finding it covered with snow late in the season and unoccupied. It was still unoccupied when checked by DFP on 10 January 1990.

Elephant Rocks Colony. According to WRF, several pairs of shags bred on Elephant Rocks in Arthur Harbor, where the species' nesting had not been previously reported: six occupied nests in 1988-89, and three in 1989-90. No banded birds were observed by him.

Winter Records. During the winters of 1955-57, Holdgate (1963) concluded that the shags on the south coast of Anvers Island made short-range migrations to open-water feeding grounds, and that variations in sea ice distribution probably had a profound influence on the winter distribution of the species. Nothing in this study proved contrary to that view. Heimark and Heimark (1984) stated that the shags remained near Palmer Station throughout the winter of 1983; 40 to 50 pairs were at nests on Cormorant Island, while another thousand or more individuals fed in Arthur Harbor on 24 July.

During WinCruise I, Pietz and Strong (1986) recorded the species between 63°30' S and 65°30' S from 25 August to 18 September 1985. At the Cormorant Island colony on 30 August, they counted 1,020 shags, virtually all of them paired. They also recorded fair numbers near the northeast coast of Anvers Island. From a flock that flew into Fournier Bay on 26 August, they collected a pair of adults; near the Melchior Islands they observed a flock of about 300 birds on 17 September.

Foraging. Bernstein and Maxson (1984) determined that during the breeding season male shags foraged approximately from 1200 to 2400, and females at other times, except during darkness, when both sexes were at the nest. Day length influenced the rhythms: timing of the males' return to the colony correlated with sunset, and timing of the females' departure correlated with sunrise. Other influencing factors were strong winds and pack ice, which, respectively, impeded flying and increased the distance to open-water feeding areas. Shags of one sex or the other often foraged singly or in groups of various sizes. Flying for-

Antarctic Blue-eyed Shag

mations frequently contained several hundred birds. Feeding flocks often dove synchronously.

Flight speed for the shags was determined by Bernstein and Maxson (1985) to be about 52 kilometers per hour airspeed when winds were below 10 kilometers per hour. During favorable periods when food was abundant, probably little time and energy were lost in flying to the feeding area, as the distance was thought to average only 10 kilometers. At the feeding areas dive times ranged from 5 seconds to 3.5 minutes; interdive times were from 2 seconds to 4.3 minutes. Combined surface swimming and diving amounted to only 8.4 percent of the time away from the colony,

Antarctic Blue-eyed Shag

whereas 91.6 percent was spent on shore resting or in low-energy activities. It was not determined whether these foraging tactics were used throughout the year. Also not determined was the maximum depth at which shags were able to dive for food. Elsewhere they dive to depths of at least 25 meters (Conroy and Twelves 1972).

Bernstein and Maxson hypothesized that foraging efficiency, through the reduction of intersexual competition, might have influenced the divergence of foraging times; once the feeding times diverged, any male that left to feed with the females also left his territory and nest unguarded. Whether this divergence also resulted in the sexual segregation of feeding niches remains a moot question, one that probably will not be resolved until it is known whether the sexes exploit different habitats and capture different species, and possibly different sizes, of fish.

Diet. Other than the observation that fish appeared to be the principal if not the only food consumed by the shags, very little information was obtained on the subject in this study. One adult male and one adult female specimen obtained by Pietz and Strong (1986) at Fournier Bay, Anvers Island, on 26 August 1985 contained only two and one fish, respectively. One of the

fish was identified to the genus *Notothenia*. More information is needed on the kinds of fish and perhaps other foods preyed on by shags, and especially needed are more studies on the cyclic behaviors and distribution of fish in these antarctic regions. Some of the best studies to date on the prey items taken by blue-eyed shags are from Schlatter and Moreno (1976) and Shaw (1981).

Molt. Bernstein and Maxson (1981) observed the plumage patterns of Antarctic Blue-eyed Shags for fourteen months and altered previous concepts on the species' molt. They discovered that the shags do not have two separate molts, but rather a nearly continuous multicycle molt sometimes referred to as the Staffelmauser. Of the 93 males and 78 females examined at the Cormorant Island colony, all but 11 showed some form of molt throughout the study including the flight feathers, which usually contained old ones among the new. The advantage attributed to this kind of arrangement is one of energy conservation when food sources are depressed. Resident shags that overwinter in the Antarctic fit these requirements.

The Cormorant Island studies showed that the heaviest molt of all feather tracts in both sexes occurred between late March and mid-April. By mid-May the nuptial crests were highly developed; these were retained until the third week in December. White dorsal patches and wing bars were not, as previously believed, replaced with dark feathers during a molt. Rather, they were replaced with white ones in all but a few individuals that showed sparse black areas amounting to only 25 percent of the patch. The Devillers and Terschuren (1978) description of the molt for the dorsal patch and wing bars simply does not apply to the antarctic population.

Since that time Rasmussen (1988a, b) made a detailed study of the South American population. In her feather-by-feather analysis, she found significant differences between subadults and adults in the step-wise molt of their flight feathers. With respect to the wing remiges, subadults have more molting feathers per wave in primaries and secondaries than do adults, but do not differ in the number of molting feathers per wing. Molt of the tail rectrices is not symmetrical in both age groups: in adults, however, the number of molting rectrices and the number of waves are correlated, whereas in subadults the number of molting feathers is not correlated, but the number of molting waves and the number of retained juvenile feathers are correlated. Rasmussen concluded that the molt of the flight and body feathers mostly takes place after breeding, with only limited molting during breeding

and in winter—a condition reminiscent of the findings at Cormorant Island. In this study it was not determined when the shag assumed a full adult plumage. Evidently, P. Shaw observed two- and three-year-olds with juvenile primaries and secondaries at the South Orkney Islands (Bernstein and Maxson 1981).

Wing-spreading. Wing-spreading, as observed commonly in the shags of more temperate climates, was not observed in this study. Bernstein and Maxson (1982b) hypothesized that it is not necessary for the Antarctic Blue-eyed Shag to spread its wings for drying; to do so would be a disadvantage for resident birds in a continuously cold climate. The species' dense inner plumage and air layer prevents icy water from contacting the skin, while the wettable surface plumage gives the advantage of being more hydrodynamic due to less buoyancy and drag during swimming. Although wing-spreading in milder climates would promote drying and heat retention, it would have the opposite effect in cold climates. Rasmussen and Humphrey (1988) studied wing-spreading in the Chilean population and concluded that their observations supported the hypothesis by Bernstein and Maxson that climatic factors outweigh the putative advantages of wing-spreading.

Eyelids and Nasal Caruncles. In conjunction with their study of the molt, Bernstein and Maxson (1981) observed the changing conditions of the eyelids and nasal caruncles in order to observe intraspecific variability within the yearly cycle. At the beginning of courtship in late September, the eyelids were a deep shade of cobalt blue, and the enlarged caruncles were a bright orange-yellow. But all of this was short-lived. By egg-laying (mid-November to mid-December), both the eyelids and the then-shrunken caruncles had faded in color, and they remained so until the following breeding season. Investigators arriving late on the antarctic breeding grounds would fail to see them at their peak condition.

Predation. Sheathbills, which frequently reside within or close to shag colonies, were the most conspicuous potential predators. Although they have a reputation for taking eggs and small chicks, in this study they were mostly observed scavenging around shag nests, frequently taking advantage of dropped food when nestlings were being fed. Skuas were also potential predators. Bernstein and Maxson attributed the death of an eight-week-old shag chick to a blow on the head by a Brown Skua. There was no evidence that gulls preyed upon anything but fish and invertebrates. Also not determined was whether Leopard Seals preyed on shag fledglings, as they did occasionally on penguins in the same waters.

Shags were kleptoparasitized by South Polar Skuas during those periods when the predators had difficulty foraging for surface fish. Casual observation of this rather unusual phenomenon prompted Maxson and Bernstein (1982) to study it in depth from 28 October 1979 to 11 March 1980, when 165 hours of scheduled observations were taken. Skuas attacked the shags as they returned from their foragings, most often several kilometers from the colony itself, and tried to induce the shags to regurgitate their fish through aerial harassment. The shag's most effective means of escape was landing on open waters; all successful chases occurred during periods of dense pack ice. Of 280 chases observed, only 13 (4.6 percent) succeeded, and most of them at times when the skua actually struck the shag in midair. Considering that the chases often covered distances of 500 meters or more, with accompanying energy expenditures, Maxson and Bernstein concluded that the skuas did not achieve much of an energy profit from the low success rate. The skuas, however, had little choice, considering the ice cover that prevented their normal foragings.

Mortality. Antarctic Blue-eyed Shag mortality appeared to be very high at times, judging by the few nesting pairs at the Cormorant Island colony in 1974-75. The shags had another particularly stressful time in 1978-79, when Bernstein (1982) witnessed a low return of breeders, high rates of nest abandonment and egg failures, and low rates of chick survival. Of the 30 pairs marked that season, 18 (30 percent) failed to return to the colony in 1979-80 and were presumed dead. Bernstein believed that low food supply was the probable cause. The widespread starvation of chicks during 1979-80 led Maxson and Bernstein (1980) to speculate that the lighter-than-normal ice cover and extremely clear water in spring forced negatively phototropic fish to depths below the diving range of shags. Evidently the shags of Cormorant Island were also one of the bird species most adversely affected by the 1989 oil spill at Arthur Harbor (Penhale 1989).

Breeding Biology. Most of the information obtained in the study on colonial shag activity was collected by Bernstein and Maxson, who camped for long periods on Cormorant Island and logged 3,198 bird-hours of time budget data from 15 January 1979 to 1 April 1979 and from 23 September 1979 to 15 March 1980. Their most revealing conclusion was that while the patterns of time allocation between the sexes differed significantly, both sexes allocated approximately equal amounts of time for reproduction.

Nest sites. Nests were sometimes isolated, but usually situated colonially at the edge of the sea on the sides or tops of low cliffs, not infrequently on isolated stacks or on piles of scree below towering cliffs. The shags tended to nest on north-facing slopes where snow disappeared early. Most observed nests were easily accessible to humans without climbing equipment, but a few were high on the sides of precipitous cliffs. The most atypical site encountered was at an abandoned Chilean station at Waterboat Point (64°49' S, 62°52' W) on the Antarctic Peninsula: on 6 January 1989, two pairs of shags brooded chicks in nests built on the tops of empty fuel drums inside a wooden shed, the floor of which was covered with nesting Gentoo Penguins. Shags often but not invariably nested close to and sometimes even among nesting penguins.

Density. From a few to as many as 500 or more pairs constituted a single isolated colony. The majority of colonies spread across Palmer Archipelago ranged between 20 and 50 pairs, often with small sub-colonies near a main colony. In dense colonial situations, neighboring nests were spaced barely far enough apart for individuals to walk unhampered among incubating birds.

Site tenacity. Banded individuals indicated that with few exceptions adults showed a strong tendency to nest annually in the same area of the colony, males at their old nests, females at other nests close by. Young hatched at the Cormorant Island colony later returned as prebreeders and still later as breeding adults, the one exception being the short dispersal to an incipient colony at Christine Island. A concerted search for banded individuals at other colonies throughout the Archipelago yielded not a single recovery.

Mate retention. A high mortality of marked shags during 1980 prevented Bernstein and Maxson (1982a) from comparing the effects of successful or unsuccessful nesting between seasons, since both members of the pair survived at only twelve previously successful nests. Of the twelve surviving pairs, as many as seven (58 percent) of them switched mates. The high incidence of mate switching was unexpected at a colony visited frequently by paired birds during nonbreeding—a behavior thought to reinforce pair-bonding. Mate change between seasons was also high (77 percent) at a colony studied by Shaw (1985b) in the South Orkney Islands. Why this should be remains a moot question.

Antarctic Blue-eyed Shag

Site defense. Males established and defended a territory until paired. Once paired, both sexes defended the nest, not only against predators but also against conspecifics, especially territory usurpers early in the season. All season long they defended against nest-material thievery by neighboring birds. Fleeing from the nest when approached by humans and other intruders was uncommon. Under less pristine conditions in South America, DFP witnessed the mass bolting of entire groups of nesting shags at the mere approach of humans from

**Antarctic Blue-eyed
Shag**

a distance. Most incubating or brooding
adults in this study held fast to their nests
and directed head-on threats at the
intruder. Threat displays included head
waving from side to side, expanded throat,
widely exposed orange gape, raised plumage and erected crest, partially opened
wings, and vocalizations amounting to
mere hissings in females and low, guttural
sounds in males (described as "aark").
When approached closely, or when handled
during banding, the birds struck out menacingly with their bills, which are capable
of inflicting painful wounds. Shags occasionally fought violently among themselves.
One fight witnessed by Bernstein and Maxson lasted eight minutes and involved as
many as four males.

Sexual displays. Intense sexual activity
at the Cormorant Island colony commenced in September and attained a peak
by the onset of egg-laying in November.
Courtship displays often occurred synchronously at all nests under observation.
Not all sexual displays attributed to other
shags or cormorants were recorded by Bernstein and Maxson (1982a), but many of
them were, as summarized below.

A. Postlanding display: Inflated head
and neck extended forward, lower than the
back. This was believed to have been a

combination of recovery after landing and submissive posture that graded rapidly into pair-bond maintenance.

B. Circle flying: Short flights away and back to the nest; function was uncertain.

C. Stepping: Rapid, high-stepping walk with bill held against the chest; most exaggerated in females. Although the function of this display was one of appeasement while walking through a colony of nesting shags, it was observed most commonly in females during the courtship period, when they had to approach nests closely in choosing mates.

D. Wing waving: Wing tips raised upward and outward, with primaries folded behind the secondaries. This was believed to have been a male advertising display that enhanced the white back patches.

E. Head wagging: Necks of both birds fully extended in one direction, often immediately followed by another extension 180° away from the first; necks usually parallel but sometimes crossed. This was primarily a nonvocal pair-bond maintenance and courtship behavior.

F. Pair-bond display: Similar to "throat-clicking" as described by Snow (1963). Both birds waved their heads rapidly back and forth, with the female holding her bill wide open and the male maintaining a slight gape; the head but not the neck was

moved in a horizontal plane. This was believed to have been another pair-bond maintenance behavior that was observed at the nest frequently before arrivals and departures, but also before, during, and immediately following copulation.

G. Allopreening: Bills directed to the head and neck regions, often close to the eyes. Multiple functions ascribed to this behavior include pair-bond maintenance, appeasement, and feather maintenance. Conceivably, it also functioned as external parasite removal, since ticks were observed on living birds and found on specimens as well.

Nest-building. Shag nests were conspicuous, compacted, truncated cones composed mostly of coarse algae cemented together with wet guano. Almost any loose objects—such as molted feathers, mummified corpses, even wooden-dowel nest markers—were incorporated into the nests. Although Shaw (1981) observed a few females gathering algae, Bernstein and Maxson noted that only males gathered the algae, which was obtained daily from the ocean floor in nearby littoral waters at midday early in the breeding season. Distance from the Cormorant Island colony to the kelp beds was approximately 0.4 kilometer; the complete trip, including flight time and dive time, averaged only 4.4 minutes. More than 95 percent of all nests examined con-

tained the alga *Desmarestia menziesii*, but *Plocamium cartilagineum* and *Gigartina skottsbergii* were also used.

Although both sexes engaged in nest construction, usually it was the female that incorporated the algae into the nest with quivering motions of the bill. Nest-building began as early as 20 September (1979) and, by early November, many nests had taken shape. It culminated during the laying-incubation stage, when females did most of the work through manipulation of the algae that had been delivered to them. Although males spent less time building during the early stages of chick rearing, they caught up with the females during the early and middle stages. Nest-building all but ceased during later stages. However, at no time did algae gathering by the males, and construction by either sex, take up more than 8 percent of their daily time budgets.

Egg-laying. When first observed on 25 November 1973, Cormorant Island nests for the most part were eggless; comparatively few had one or two eggs, and in only two cases were there three. Conditions were similar on 20 November 1975, when only a single nest had a three-egg clutch. Early egg dates were also recorded by Heimark and Heimark (1984): on 16 November 1983 they estimated that 80 percent of the 378 or more nests had at least one egg, and in a "few" cases there were three. Laying of these three-egg clutches probably started no later than 11 November, judging by the 2.5-day interval between layings as determined by Shaw (1981).

Bernstein and Maxson (1984) stated that the shags were asynchronous breeders, in some cases showing as much as a six-week difference between laying dates. DFP noted fresh eggs at the Cormorant Island colony in 1985 as late as 13 January, and some fairly small chicks still being brooded at the Joubin Island colony in 1979 as late as 5 February. Not determined was whether any of these late layings were replacements for lost clutches.

Clutch size. Clutch size was generally two or three eggs. One four-egg clutch was noted at the Cormorant Island colony, and another at the Joubin Island colony. The mean clutch size was 2.5 for 811 Cormorant Island nests examined by Bernstein and Maxson (1985). The mean egg weight for 13 eggs from these nests was 59.1 grams.

Incubation. Both sexes incubated, but, according to Bernstein and Maxson (1985), males incubated more than females. Neither sex spent much time at the nest when not incubating. The period of incubation was not determined.

Hatching. DFP recorded young at least 5 days old at Cormorant Island on 13 December 1984, indicating a hatching date possibly as early as 8 December. Presumably young from late clutches hatched during February, as was almost certainly the case with late-fledging individuals.

Brooding. Ricklefs (1982) found that, as might well be expected of naked hatchlings in a polar environment, the Cormorant Island shags had poorly developed homeothermic abilities at hatching and were the last of four other local species to develop effective control of body temperature. Both sexes brooded, but, according to Bernstein and Maxson (1985), males brooded the most, especially during the early chick-rearing stage, while females rested by the nest. As the season progressed females spent less time by the nest and more time foraging. Males and females fed the chicks with equal frequency.

Prefledging. During the late rearing stage, females increased nest-centered activities. Although they foraged more than males, time budgets were approximately equal for the two sexes. Chicks were brooded until they were about 24 days old, but they were still guarded at the nest for at least another three weeks.

Fledging. According to Bernstein and Maxson (1984), the ulna wing bone in chicks stopped growing at day 45, at about

the time they were left unguarded. Chicks flew strongly between 55 and 60 days of age, and this figure falls fairly close to the mean fledging period of 65 days stated by Shaw (1981). The earliest date recorded for fledged birds at Cormorant Island was 4 February (1979), the latest for unfledged birds was 4 March (1980). In 1989, at Cuverville Island near the Antarctic Peninsula, DFP noted weak flying young as early as 5 February, and young still not fledged as late as 1 March.

Postfledging. Fledglings tended to flock not only at the colony but also nearby on ice floes, rocky points, or in the water, where they swam, bathed, and foraged, sometimes in company with adults. According to Bernstein and Maxson (1985), older fledglings typically flew from the colony during the morning and returned in all-juvenile groups about an hour before the adults returned. The young were then fed, usually in late afternoon. It was not determined how long the parents fed their fledglings. Snow (1960) observed that 100-day-old Shags (*Phalacrocorax aristotelis*) were still being fed by adults.

Breeding success. Although shag breeding success fluctuated considerably from season to season, the two-year study at Cormorant Island by Bernstein and Maxson showed high mortality. In 1979-80, about 40 percent of the mortality was due to egg infertility, nest abandonment, or predation, and about 35 percent to chick mortality; only 25 percent of the eggs produced fledglings. Productivity in that year was 75 percent lower than in the previous year. Reports from other stations in the region indicated similar nesting failures in 1979-80. The species fared much better at other times. High fledging success was evident at many colonies visited in the region by DFP in 1988-89.

Synopsis of annual cycle. Antarctic Blue-eyed Shags resided at the nesting colonies throughout the year, even when their nest sites were covered with winter snow. Males staked out territories by September, when courtship was well under way. Site tenacity was strong, but mate retention was weak, despite the year-long, repeated returns to the colony. Although the sexes had distinctly different foraging and nesting behaviors, each contributed nearly equal amounts of time and energy to the reproductive effort. Breeding success was closely linked to the availability of fish prey. Egg-laying, hatching, and fledging were highly asynchronous, resulting in a lengthy breeding season. The period from the laying of the first eggs to the fledging of the last young spanned a minimum 113 days. Although fledglings flocked early, they returned daily to the colony, where they were fed by adults for an undetermined period. At least some young returned to the natal nesting colony, where they commenced breeding as early as four years of age. Dispersal to incipient colonies close by occurred, but it was not determined whether any of the fledglings settled in distant colonies.

Family of Waterfowl

Several species of waterfowl are resident breeders on islands in the Sub-Antarctic, but only vagrants have been reported in the vicinity of the Antarctic Peninsula. Although one species of swan and two species of ducks were recorded for Palmer Archipelago, their occurrences there varied between seasons and were highly unpredictable. All these birds probably flew across the Drake Passage from their native grounds in South America.

Yellow-billed Pintail.

Black-necked Swan
(*Cygnus melanocoryphus*)

Overall Breeding Distribution. Breeds in South America, including Chile, Argentina, and the Falkland Islands.

Current Status in Palmer Archipelago. Rare vagrant.

Summer Records. The first sighting of a Black-necked Swan for the Antarctic Peninsula region occurred during the austral summer of 1916-17 and concerned an emaciated individual captured in "Charlotte Channel, South Shetland Islands," as reported by Bennett (1922). According to Watson (1975), the locality was really Charlotte Bay (64°33' S, 61°39' W), Ant-

Table 15. Black-necked Swan (*Cygnus melanocoryphus*) sightings at South Shetland Islands, Palmer Archipelago, and Antarctic Peninsula during 1988-89

No.	Age	Date	Locality	Observer(s)
several		Nov. 88	Palmer Station (64°46′ S, 64°03′ W)	Palmer personnel (pers. comm.)
5	adults	27 Dec. 88	Palmer Station	WRF
2	adults	04 Jan. 89	Palmer Station	WRF
1	adult	07 Jan. 89	South Shetlands, Nelson Island (62°19′ S, 59°15′ W)	M. Favero (pers. comm.)
1	adult	20 Jan. 89	Palmer Station	WRF
1	adult	29 Jan. 89	South Shetlands, Deception Island (62°57′ S, 60°38′ W)	J. Miller (pers. comm.)
1	adult male	06 Feb. 89	Wiencke Island (64°49′ S, 63°30′ W)	DFP, R. Schlatter, and W. Sladen
4	adults	23 Feb. 89	Palmer Station	WRF
1	adult	03 Mar. 89	Antarctic Peninsula, Paradise Bay (64°54′ S, 62°52′ W)	E. Carriazo (pers. comm.)

arctic Peninsula. Parmelee and Fraser (1989) reported on multiple sightings of Black-necked Swans observed at the South Shetland Islands, Palmer Archipelago, and the Antarctic Peninsula during 1988-89. Table 15 shows the localities and dates of the sightings. The five adults observed near Palmer Station on 27 December 1988 represented the largest concentration of the birds seen at one time. Conceivably, individuals of this group later dispersed, but it is equally possible that one or more individuals crossed the Drake Passage independently. All the swans appeared to be healthy. According to R. Schlatter (personal communication), swans probably survive fairly well in these coastal waters until the winter freeze. Fraser saw one feeding on a swarm of krill near Palmer Station on 20 January 1989.

Yellow-billed Pintail
(*Anas georgica*)

Overall Breeding Distribution. Breeds chiefly in South America, but also in the transitional and cold Sub-Antarctic. Subspecies *A. g. spinicauda* breeds from Peru and southern Brazil to Tierra del Fuego and the Falkland Islands. *A. g. georgica* is confined to South Georgia.

Current Status in Palmer Archipelago. Uncommon vagrant. Several sightings to date, all within the Palmer study area.

Summer Records. One was noted at Breaker Island on 18 January 1975 (Parmelee et al. 1977b). During 1979-80, NPB noted three at Bonaparte Point on 31 October and collected a male there on 1 November. Thereafter, singles were sighted one time or another by DFP, SJM, NPB, and G-AM at Humble Island on 25 November, Cormorant Island on 2 December, Limitrophe Island on 20 January, and Hellerman Rocks on 9 March. The scattered remains of one were found that same season by SJM at Hermit Island on 29 February.

Subspecies. The male specimen collected on 1 November 1980 was noticeably paler and larger (weight, 553.0 grams; culmen, 44.0 millimeters; right wing chord, 240.0; tail, 120.0; tarsus, 43.5) than specimens examined from South Georgia, thus referable to the South American race *Anas georgica spinicauda* rather than *A. g. georgica*.

Chiloe Wigeon
(*Anas sibilatrix*)

Overall Breeding Distribution. Breeds chiefly in Chile and Argentina, and sparingly in the Falklands.

Current Status in Palmer Archipelago. Rare vagrant. Four sightings to date, all within the Palmer study area in 1980.

Summer Records. A single drake (conceivably the same individual each time) was sighted at one time or another by DFP, SJM, NPB, PJP, and G-AM at Hero Inlet on 25 February, Arthur Harbor on 27 February, Humble Island on 3 March, and Norsel Point on 8 March.

Subfamily of Phalaropes

halaropes are shorebirds that migrate from the Northern Hemisphere to far southern areas in winter, but they are rarely encountered as far south as Antarctica, where they can be rightly called extreme vagrants. Watson (1975) lists only two records for that region, based on a specimen of Wilson's Phalarope (*Phalaropus tricolor*) found dead at Fossil Bluff (71°20' S, 68°17' W), Alexander Island, in mid-October 1968, and on a specimen of Red Phalarope (*P. fulicaria*) collected near Palmer Station in mid-January 1970 before this study was under way.

Closely related sandpipers of the subfamily Scolopacinae are vagrants more likely to be seen in Palmer Archipelago. J. Jolmen (personal communication) noted a sandpiperlike bird in the Palmer study area at Stepping Stones on 3 January 1984. His description fit that of a White-rumped Sandpiper (*Calidris fuscicollis*), a likely candidate, but the bird disappeared before its identity could be confirmed.

Pair of Red Phalaropes in breeding plumage at the 76th parallel on Bathurst Island, Canadian Arctic. The smaller and duller male is at the left. Photographed 20 June 1970.

Red Phalarope
(*Phalaropus fulicaria*)

Overall Breeding Distribution. Circumpolar in the Arctic, including Alaska, Canada, Greenland, Iceland, and northern Siberia.

Current Status in Palmer Archipelago. Rare vagrant: one record.

Summer Record. A single male in breeding condition (reddish plumage, bright yellow bill, enlarged testes) was collected near Palmer Station on Humble Island on 12 January 1970 (Risebrough et al. 1976). The plumage condition is noteworthy because at that time of year the species normally has a dull winter appearance (gray plumage, blackish bill) that is strikingly different from the bright-colored one seen on its breeding ground in the Arctic. The specimen is in the National Museum of Natural History, Smithsonian Institution, Washington, D.C.

Family of Sheathbills

Greater Sheathbill

Sheathbills are peculiar birds that are thought by some ornithologists to be a connecting link between the shorebirds and jaegers and gulls. Their most outstanding anatomical feature is a compressed, conical bill with a horny sheath that extends over the base of the upper bill and covers the nasal openings. Fleshy wattles occur about the bill and face. Sheathbills also have spurs at the joints of the wings and, except for rudimentary webs between the front toes, are essentially webless—a rare condition in Antarctica. Superficially, sheathbills look somewhat like white doves; they run about like domestic fowl, although they are strong flyers when induced to flight. Only two species constitute the family, and both are confined mostly to the Southern Ocean.

Of the two species, the Greater Sheathbill (*Chionis alba*) occupies the Scotia Sea and Antarctic Peninsula regions, including Palmer Archipelago, where it was conspicuous at many penguin and shag colonies. Although its primary nesting sites were concealed crevices in natural situations, it read-ily adapted to open sheds, empty fuel drums, almost anything introduced and later abandoned by humans. Although some sheathbills remained throughout winter at Palmer Station, others moved about locally, or migrated between there and the South Shetland Islands. It was not determined whether any of these sheathbills migrated across the Drake Passage to Chile, Argentina, or the Falkland Islands, where nonbreeding individuals and flocks occur regularly.

So few sheathbills nested in the vicinity of Palmer Station that only limited observations were obtained on them during the summer months. DRN banded and observed sheathbills that visited the station during the winter of 1975. Likewise, Shaw (1986) banded and observed sheathbills during the winter of 1980 at the British Antarctic Survey Station on Signy Island, South Orkney Islands, and Peter et al. (1988a, b) banded and observed them during the winters of 1984 and 1985 at the Bellingshausen Station on King George Island, South Shetland Islands. Jones (1963) studied nesting Greater Sheathbills extensively in the South Orkney Islands, and Burger (1980) made a comprehensive study of its Indian Ocean congener, the Lesser Sheathbill (*C. minor*).

Greater Sheathbill
(*Chionis alba*)

Overall Breeding Distribution. Breeds in the Maritime Antarctic and transitional Sub-Antarctic, including South Georgia, South Orkney, South Shetlands, and the Antarctic Peninsula south to about the sixty-fifth parallel.

Current Status in Palmer Archipelago. Common year-round resident. Breeding distribution was linked closely to penguin and shag colonies. Migrations were poorly understood within and beyond Palmer Archipelago.

Banding. A total of 143 Greater Sheathbills were banded within the Palmer study area.

Adults/Subadults. DRN banded one nesting pair at Cormorant Island on 17 January 1975. Before the closing of the dump at Palmer Station that same year, he banded the following: May, 37 birds; June, 29; July, 13; August, 6; September, 17; October, 10; November, 1. Following the closing of the dump, WRF banded only 14 birds there from 25 July to 5 September 1976, BMG a mere 5 birds from 22 July to 5 September 1977, and NPB banded not one in 1978. At Cormorant Island, SJM and NPB banded one pair and a single adult, respectively, on 22 and 30 January 1978.

To date there have been two distant recoveries of sheathbills that were banded at Palmer Station. One banded by DRN on 12 May 1975 was recovered about 110 kilometers from Palmer on 19 December 1985 at Metchnikoff Point, Brabant Island, by a British expedition party under J. R. Furse. Another banded by DRN on 16 September 1975 was recovered in the South Shetland Islands on 9 October 1981 by W. Z. Trivelpiece at Admiralty Bay, King George Island, about 400 kilometers from Palmer. Possibly these birds had wintered south of their breeding areas.

To date there has been but one Palmer recovery of a sheathbill that had been banded elsewhere: an adult banded by the British Antarctic Survey on 29 August 1967 at Galindez Island (65°15' S, 64°15' W), Argentine Islands, was recovered by DRN on 17 January 1975 at Cormorant Island, where it had nested during 1974-75 but not thereafter.

Nestlings. At Cormorant Island, SJM and NPB banded two siblings at one nest on 22 January 1979, and two siblings at each of two nests on 10 February 1980. None has been recovered to date.

Summer Records. Sheathbill distribution nearly coincided with the shag colonies throughout Palmer Archipelago. The species also favored Chinstrap and Gentoo penguin colonies, but for some unknown reason not the Adélie Penguin breeding colonies. Within the Palmer study area, one to two pairs of sheathbills nested annually at Cormorant Island, where a few nonbreeders also occurred. Outside the study area, the closest nesting sheathbills occupied a shag colony at the Joubin Islands (two to three pairs), and a shag-Gentoo colony at Port Lockroy, Wiencke Island (two to four pairs).

Winter Records. DRN noted that sheathbill attendance fluctuated throughout the winter at Palmer, where peak numbers occurred several times monthly during May, June, July, August, and September. The highest peaks occurred during 11-17

July, when as many as 30 individuals visited the station dump. Numbers dropped noticeably in October and virtually ceased by November.

Of the 115 sheathbills banded by DRN from 3 May through 1 November 1975, 23 were marked with color bands or tape. Observations of these birds indicated that the sheathbills moved freely not only about the islands within the study area, but also in and out of the area. Of the 23 marked birds, 6 were not seen again and 9 returned to the station repeatedly. The remaining 8 were recorded only once or twice following their initial capture, though one to five months elapsed between visits by some of them.

In the South Shetland Islands, Peter et al. (1988b) stated that the number of sheathbills increased from April to late May and was stable in June and July. The birds departed for unknown nesting grounds following a second peak in September. Numbers of first-year birds decreased substantially by midsummer. Adults banded in April or May showed the highest site fidelity, and the percentage of wintering males seemed to be higher than that of females.

In the South Orkney Islands, Shaw (1986) concentrated on dominance behavior related to the age and bill size of his wintering flock. The dominant birds were known to be over six and seven years of age and possessed the largest bills. They also resumed feeding more quickly than the subordinates they displaced. Despite the advantage of a higher peck rate, they did not show a higher survival rate than their subordinates—an interesting observation that merits further study.

Diet. The sheathbill's summer foraging grounds were shag and penguin colonies, where they preyed on eggs and small chicks, but mostly they pirated bits of food dropped by the shags and penguins whenever they attempted to feed their young. In addition to open station dumps, they frequented the winter roosts of shags on Cormorant Island and those of Adélie Penguins on Torgersen Island, where, according to DRN, they fed on fresh shag and penguin excrement, as well as on seal excrement and carcasses.

Mortality. Mud and ice accumulation on the bodies of sheathbills impaired the birds' mobility. During the winter of 1975, DRN observed a sheathbill with "mud balls" clinging to its primaries, rendering the bird flightless; although this individual survived, its left wing was visibly damaged. During the winter of 1976, WRF observed ice the size of golf balls on the feet of several sheathbills. The accumulation severely hampered their movements and resulted in

Upright display in defense of territory.

at least one death. Not determined was how these ice and mud balls formed in the first place.

Mud was detrimental to sheathbills during breeding as well. Unattended, very cold eggs at two Cormorant Island nests were covered with thick layers of mud on 24 December 1983. Two abandoned eggs at one of the sites contained no embryos when examined on 2 January, and the single, long-abandoned egg at the other site eventually disappeared. No additional eggs were laid that season. The partial remains of sheathbills were found occasionally in the study area, including the carcass of one at a skua's nest.

Breeding Biology.

Nest sites. Nests occurred from near sea level to the tops of high ridges on islands and headlands, invariably close to their prey species. Usually they were built near the innermost limits of small caves, fissures, scree, and the like, often beyond human reach. Some, however, were tucked in shallow crevices open to the sky, presumably because of the scarcity of prime sites where pairs were especially abundant. Some nests were in burrows excavated in shag guano beneath large boulders; conceivably, the sheathbills were the only local vertebrates capable of constructing such burrows up to 2 meters long. They also nested in almost any hidden cranny left behind by humans. Their adaptiveness to artificial sites was especially apparent at the abandoned British station at Port Lockroy, Wiencke Island.

Nests. Sheathbill nests were usually well-constructed affairs made of any transportable item, including penguin feathers and egg shells, bones, pebbles, limpet shells, guano, moss, lichens, grass, seaweeds, and sheathbill egg membrane and shell remains. Old nests were refurbished, often annually.

Density. One to three pairs were encountered at most of the shag, Chinstrap, and Gentoo penguin colonies visited in Palmer Archipelago. A notable exception was the large Chinstrap colony on Brabant Island at Metchnikoff Point, where a dozen or more sheathbill pairs resided in 1983-84. Large concentrations were also noted at the South Shetland Islands (Nelson and Half Moon islands) and Antarctic Peninsula (Waterboat Point).

Site fidelity. Site 1 on Cormorant Island was occupied by a banded pair of sheathbills that disappeared after the 1974-75 season. It was next occupied in 1975-76 by an individual that had been banded by DRN at Palmer Station on 15 September 1975; the marked bird was still nesting on Cormorant Island with an unbanded mate when last seen in 1984-85—nearly a decade later.

Site 2 was observed for the first time in 1977-78, but the pair was not banded there until the following season. On 1 January 1983, parasitologist Hoberg (1983) collected three male and two female sheathbills at Cormorant Island, including the banded male from Site 2, which has remained unoccupied since that time.

Site defense. Both sexes of sheathbills confronted territorial intruders with agonistic displays that included wiping the bill vigorously on the ground followed by facing the intruder head-on with lowered head and forward-pointing bill while pumping the tail and vocalizing loudly. These threat postures were described as the "forward-oblique" display by Jones (1963), or simply the "forward" display by Burger (1980). Another aggressive display previously described and noted commonly in this study was the "upright" display, in which the sheathbill stood in an extended upright posture with closed or slightly opened wings, at times squarely facing an adversary that had assumed a similar pose. "Chases" were frequent and occasionally ended in fierce combat, especially between males in the presence of a mate.

Incubating or brooding sheathbills usually ran from their nests when approached by humans, but then scampered close by while vocalizing in a highly agitated manner before quickly returning to their eggs or chicks at the first opportunity. A few that were reluctant to leave the nest struck out with their bills menacingly. Paired individuals frequently bowed to each other while uttering sharp, staccato calls before entering the nest cavity, or during other activities. This display was thought not to be agonistic; rather, it appeared to function to maintain the pair-bond. It was called the "bowing ceremony" by Jones (1963) and the "bob call" by Burger (1980). Certainly it was the most conspicuous display noted in Palmer Archipelago.

Breeding chronology. Although marked individuals visited their Cormorant Island breeding ground throughout winter, the earliest date recorded for nest site occupancy was 15 October 1979. That day NPB saw the pair display at Site 1, where they engaged in bob calling, while the pair at Site 2 chased intruders. The first egg at Site 2 had been laid by 29 November, followed by the second on 30 November—the earliest precise date recorded for egg-laying in this study. A third egg appeared some time later. A single egg was first observed at Site 1 on 3 December, but for some reason it was abandoned and the pair then occupied a fresh burrow 10 meters away and produced a second clutch of three eggs that escaped notice for some days. Although the hatching dates were not determined, NPB noted that a chick at Site 2 was nearly fledged by 26 February, and that it and its one surviving sibling were fully fledged by 4 March. Evidently egg-laying by these same pairs had started even earlier during the previous 1978-79 season, since chicks from both sites fledged by 23 February.

Early laying was also recorded at Port Lockroy, Wiencke Island, where chicks possibly a week old were noted by DFP as early as 5 January 1989. Not far away at Waterboat Point, Antarctic Peninsula, five newly hatched chicks were noted on 6 January. Allowing for a minimum incubation period of twenty-eight days (Watson 1975), eggs at Port Lockroy must have been laid on or slightly before 1 December, those at Waterboat Point hardly more than a week later.

Judging by most nestings observed in this study, egg-laying probably peaked during mid- to late December. Within Palmer Archipelago, fresh eggs or those with small embryos were recorded by DFP at one nest on Brabant Island at Metchnikoff Point on 29 December 1983; also, there were two nests at Cormorant Island on 13 December 1984, one at Port Lockroy on 29 December 1984, one at Joubin Islands on 31 December 1984, and one near Gerlache Point, Anvers Island, on 2 January 1985. DFP recorded the following at the South Shetland Islands: one nest at False Bay, Livingston Island; also six nests at Harmony Cove, Nelson Island, on 13 December 1973, and two at Walker Bay, Elephant Island, on 19 December 1974; on the Antarctic Peninsula, one nest at Hope Bay on 20 December 1974, and one at Spigot Point on 31 December 1983.

Clutch size. Clutch sizes of one to three eggs have been recorded. Where the clutch was thought to have been completed, two nests had single eggs, seven had two eggs, and nine had three eggs.

Synopsis of annual cycle. Greater Sheathbills moved freely about Palmer Archipelago during winter, settling temporarily where they were fed artificially. Although migration or dispersal northward to the South Shetland Islands was documented, it was not determined whether any of these birds migrated to South America. Sheathbills resided within or close to penguin and shag colonies, where they scavenged food from roosting birds in winter and from nesting birds in summer. Pairs occupied nesting cavities or burrows from mid-October, including both natural sites and artificial ones related to human activity. Egg-laying commenced in late November, but most eggs were probably laid during mid- to late December, exceptionally later. Young from early nestings fledged during 23 February-4 March, those from late nestings probably two to four weeks later. At one site the period from egg-laying to fledging was approximately ninety-five days. Avian predators had little impact on sheathbill mortality, but contributing factors were the accumulation of mud and ice on the birds' bodies in winter and mud on their eggs during breeding.

Subfamily of Skuas and Jaegers

Large skuas of the genus *Catharacta* commonly breed in the Southern Hemisphere, but much speculation exists on how many species are represented. Several morphologically distinct populations are often referred to as *hamiltoni* of the Tristan da Cuna group in the South Atlantic, *antarctica* of the Falkland Islands and Argentina, *chilensis* of Chile and Argentina, *lonnbergi* of the Sub-Antarctic, and *maccormicki* of the Antarctic (see Figure 10). In this study the latter two are considered distinct species: *Catharacta lonnbergi*, Brown Skua (also called Sub-Antarctic Skua), and *Catharacta maccormicki*, South Polar Skua (also called McCormick's Skua).

Although the Brown and South Polar skuas are separated over much of their vast, circumpolar breeding ranges, some pairs nest side by side within narrow zones of overlap, including Palmer Archipelago. One of the attractive features of the Palmer study area is the opportunity to study and compare the two skuas breeding under identical environmental conditions. In addition,

mixed matings occur there, resulting in hybrids of special interest. The genus *Catharacta* is poorly represented in the northern latitudes by the single member *skua*, confined mostly to Iceland and islands north of Scotland. The view is widely held that it derived from one of the southern skuas, with *hamiltoni* being a possible candidate because of its morphological similarities and close proximity to the North Atlantic. Perhaps a more plausible candidate is the South Polar Skua, which migrates north all the way to Greenland, as demonstrated by the banding program in this study.

Other members of the Stercorariinae are noticeably smaller and also differ in having immature plumages that are barred. Because of these features they are usually classified as a separate genus, *Stercorarius*. All three species, *S. pomarinus*, *S. parasiticus*, and *S. longicaudus*, breed only in far northern regions, but commonly migrate in winter to the oceans of the Southern Hemisphere. New World ornithologists refer to them as jaegers, but Old World ornithologists retain the vernacular name skua for

Displaying "Long Call" South Polar Skua.

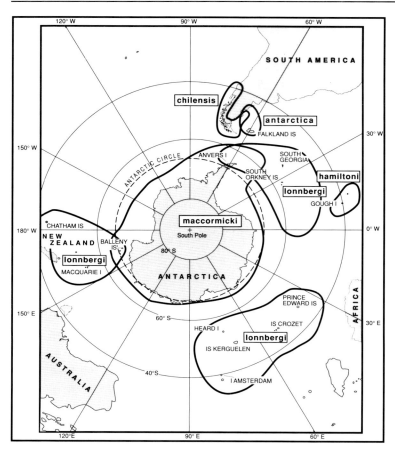

Figure 10. Approximate breeding distribution of southern *Catharacta*. Note the overlap in the distribution of *maccormicki* and *lonnbergi* in the region of the Antarctic Peninsula, including Palmer Archipelago (from Pietz 1984).

these birds as well. Both Pomarine Jaegers (*S. pomarinus*) and Parastic Jaegers (*S. parasiticus*) are vagrants in the Southern Ocean. Sightings of Pomarine Jaegers south of Anvers Island merit placing the species in the hypothetical list for Palmer Archipelago. According to Sladen (1954), an adult was observed at the Léonie Islands (67°37' S, 68°22' W) on 22 February 1937, and one at the Argentine Islands (65°15' S, 64°16' W) not far from Palmer Station on 9 February 1953.

South Polar Skua
(*Catharacta maccormicki*)

Overall Breeding Distribution. Circumpolar. Breeds chiefly on or near the main continent and Antarctic Peninsula, but also on the Balleny, South Shetland, and South Orkney islands.

Current Status in Palmer Archipelago. South Polar Skuas are abundant summer residents in Palmer Archipelago, where they occur commonly at sea and are prone to follow ships, frequently alighting on decks or other structures. Their numbers far exceed those of Brown Skuas in southern parts of the archipelago and along the Antarctic

Peninsula. A reversal in numbers occurs northward from there, being particularly noticeable in the South Shetland Islands. where Brown Skuas outnumber South Polar Skuas. Some evidence exists that the South Polar Skuas are expanding their breeding range northward; a small number of pairs currently nest at the South Orkney Islands, where the species had not been recorded until recently (Hemmings 1984; Roots 1988). Although this opportunistic species is renowned for preying on birds, in Palmer Archipelago it forages chiefly at sea for fish and krill during the breeding season, especially in the presence of penguin-dependent Brown Skuas. South Polar Skuas leave the archipelago following breeding and migrate long distances along the Atlantic and Pacific coasts of the Americas. In this study, most observations on South Polar Skuas were by Pietz and Maxson (1980), Pietz (1982, 1984, 1985, 1986, 1987), Neilson (1983), and Parmelee (1985). Physiological studies on this and other species were conducted at Palmer Station by Murrish and Tirrell (1981), Ricklefs (1982), and Ricklefs and Matthew (1983). More recently, Eppley and Rubega (1989) studied skua chick survival following the 1989 oil spill in Arthur Harbor.

Banding. A total of 2,345 South Polar Skuas were banded within the Palmer study area, and an additional 33 were banded elsewhere along the south coast of Anvers Island.

Adults. A total of 1,024 adults of uncertain age were banded within the study area, the majority on Litchfield Island (236 birds), Christine Island (219), and Shortcut Island (139). An additional 5 were banded on Dream Island. Most adults were caught at their nest sites: the best-monitored ones nested on Bonaparte Point (79 birds). During the initial study in 1974-75, DRN color banded many of the 201 adults he handled that season. Relatively few were color banded thereafter, and none during 1978-79 (219 birds) or 1983-84 (478). To date there have been no long-distance recoveries of any of these Palmer-banded adults.

Nestlings. There were 1,321 nestlings banded within the study area, and an additional 28 along the south coast of Anvers Island. Banding returns show that upon leaving their natal grounds, the juveniles traveled far and wide, notably across the Drake Passage, then northward along both coasts of the Americas (Figure 11, Table 16). Highlights of these recoveries included the following:

A. The recovery of a Palmer-banded nestling in western Greenland was good evidence for bipolar migration.

RV *Hero* docked at U.S. Palmer Station overlooking Arthur Harbor. Banded South Polar Skua from Bonaparte Point (left background) awaiting food scraps on station's balcony. Photographed 11 January 1978.

B. Migration was rapid in some cases. Following departure from early to mid-April, one individual reached Oregon by 9 July, another was as far north as Greenland by 31 July.

C. Although the recovery locations exceeded expectations, the recovery rates and dates were predictable (Tables 17, 18). The majority of recoveries (77 percent) occurred during the first year, a notable exception being one in the sixth year.

D. Two individuals were twice recovered the same year at different localities in Brazil. Another was recovered near the same locality in Brazil when three and a half and five years old.

E. Not all recoveries were coastal; a notable exception was one 400 kilometers inland, at one of Brazil's great rivers.

F. The recovery in the Gulf of California added a new species to the known Mexican avifauna.

NORTH
AMERICA

SOUTH
AMERICA

Figure 11. Long-distance returns of South Polar Skuas (*Catharacta maccormicki*) banded as nestlings near Palmer Station.

A number of these banded nestlings later bred at or near their natal areas. Of the 20 marked returnees observed during the 1983-85 seasons, 12 (60 percent) nested at the very islands or peninsulas where they had hatched; all others returned to sites close by. None mated with a known-age bird. Although both sexes were represented in the sample, insufficient numbers had been sexed to give a meaningful sex ratio among returnees. Since many banded pre-breeders were present in 1984-85, the opportunity exists for obtaining a larger sample.

Few foreign-banded South Polar Skuas have been reported for Palmer Archipelago, although one of unknown age and sex probably migrated through the region. No. 617-52019 was banded at Refugio Pas De Los Andes, Argentina, during the "winter of 1957" by C. W. Ekland, and recovered twenty-two years later far south of Palmer Station at Avian Island (67°46' S, 68°54' W) on 6 March 1979. No. 647-27144, of unknown age and sex, was banded on 5 March 1961 at the Chilean base, Gonzales Videla, Antarctic Peninsula, and recovered not far from there on Litchfield Island near Palmer Station by DRN on 12 January 1975.

Summer Records. Although South Polar Skuas were observed frequently at sea throughout much of Palmer Archipelago,

most nestings were recorded within the Palmer study area, where as many as 600 or more pairs bred some years. Nestings recorded beyond the study area are summarized in Table 19.

Arrival. The earliest arrival recorded was 20 October 1975. That season, Neilson (1983) observed the return of 33 marked individuals (17 males, 16 females) to their nesting sites on Bonaparte Point—a feat not since duplicated. The birds had been gone for a winter period that averaged 216 days. Most returned between 1 and 27 November, the average date being 12 November. Males arrived, on average, one week earlier than females (Figure 12). In all cases females arrived after their mates.

Departure. Neilson (1983) recorded the departures of 45 breeding adults at Bonaparte Point in 1975 during 25 March-22 April (Figure 13). Unlike the fledglings, the adults left the breeding grounds individually. Females left first, on average, only four days after their young. Males continued to defend their territories until shortly before departure. Females left, on average, by 4 April, males by 12 April. In 1983, Heimark and Heimark (1984) last

Table 16. Long-distance recoveries of South Polar Skuas (*Catharacta maccormicki*) banded as nestlings near Palmer Station

Band No.	Banding date	Banding locality	Recovery date	Recovery locality
877-34210 (1)	16 Jan. 75	Litchfield Island	14 Sept. 75	Mexico, Sonora, Gulf of California 31°41′ N, 114°30′ W
877-34271 (2)	20 Jan. 75	Shortcut Island	31 July 75	Greenland, Godthab 64°10′ N, 51°40′ W
877-36534 (3)	14 Jan. 76	Bonaparte Point	18 Dec. 76	Brazil, Paco do Lumiar 02°20′ S, 44°20′ W
877-36580 (4)	12 Feb. 76	Humble Island	winter 76	Brazil, Iati 09°00′ S, 36°50′ W
877-36900 (5)	17 Feb. 77	Norsel Point	10 Jan. 78	Brazil, Ilha Santana 02°10′ S, 43°40′ W
877-36723 (6)	23 Feb. 77	Limitrophe Island	30 Oct. 80	Brazil, NR Macau 05°00′ S, 36°20′ W
(6)			26 Apr. 82	Brazil, Macau 05°00′ S, 36°30′ W
1057-10116 (7)	11 Feb. 79	Eichorst Island	27 June 79	Brazil, Recife 08°00′ S, 34°50′ W
(7a)			summer 79	Brazil, Pernambuco Shore 08°90′ S, 35°90′ W
1057-10377 (8)	22 Feb. 80	Litchfield Island	09 July 80	Oregon, Bullards Beach 43°00′ N, 124°20′ W
1067-25601 (9)	06 Feb. 84	Humble Island	28 Apr. 84	Brazil, Riacho Doce 09°30′ S, 35°30′ W
1067-25838 (10)	23 Feb. 84	Litchfield Island	14 Sept. 84	Brazil, Rio São Francisco 09°20′ S, 40°30′ W
(10a)			09 Oct. 84	Brazil, Olinda Pernambuco 07°50′ S, 34°50′ W

Table 17. Long-distance recovery rates of South Polar Skuas (*Catharacta maccormicki*) banded as nestlings near Palmer Station

Year banded	No. banded	Recovered No.	(%)	Year(s) recovered
1975	161	2	(1.24)	1975
1976	42	2	(4.76)	1976
1977	287	2	(0.70)	1978, 1980, 1982[a]
1978	0			
1979	51	1	(1.96)	1979
1980	277	1	(0.36)	1980
1981	76	0	(0)	
1984	384	2	(0.52)	1984
1985	71			
Total	1,349	10	(0.74)	

[a] Same individual recovered twice.

Table 18. Long-distance recovery dates of South Polar Skuas (*Catharacta maccormicki*) banded as nestlings near Palmer Station

Year	Jan.	Feb.	Mar.	Apr.	May	June	July	Aug.	Sep.	Oct.	Nov.	Dec.	Unknown
First				1		1	2		2	1		1	2
Second	1												
Third													
Fourth										1			
Fifth													
Sixth		1											

Table 19. Nesting localities of South Polar Skuas (Catharacta maccormicki) recorded outside of the Palmer study area during 1975-85

Date		Locality	Nest(s) (contents)	Observer(s)
16 Jan.	75	Joubin Islands	1 (1 chick)	DFP
05 Feb.	79	Anvers Island (64°42′ S, 64°18′ W)	2 (2 chicks)	DFP, SJM, NPB
05 Feb.	79	Dream Island	1 (1 chick)	DFP, SJM, NPB
27 Dec.	83	Anvers Island (64°20′ S, 53°41′ W)	1 (1 egg)	V. Komarcova (pers. comm.)
29 Dec.	83	Brabant Island (64°15′ S, 62°32′ W)	1 (2 eggs)	CCR
29 Dec.	83	Brabant Island, near Astrolabe Needle	1 (1 egg), 1 (2 eggs)	DFP, CCR
07 Jan.	84	Dream Island	3 (2 eggs), 1 (1 egg), 2 (2 chicks)	DFP, CCR
12 Jan.	84	Joubin Islands	4 (2 eggs)	DFP, CCR
17 Dec.	84	Anvers Island, Biscoe Point	9 (2 eggs), 2 (1 egg)	DFP
30 Dec.	84	Wiencke Island, Port Lockroy	1 (1 egg)	DFP, JMP
05 Jan.	85	Dream Island	2 (2 eggs), 1 (egg & chick), 1 (2 chicks)	DFP
28 Jan.	85	Anvers Island, Biscoe Point	Estimated 40 pairs nesting	DFP

recorded the species in the vicinity of Palmer Station on 24 April. To date there are no confirmed records for May.

Foraging Behavior. When breeding in the vicinity of penguin-dependent Brown Skuas, the South Polar Skuas foraged mostly at sea. Their breeding success largely depended on their ability to obtain krill and fish, particularly the latter, which appeared to be vitally important to chick survival. Factors influencing their foraging at sea were ice cover, storms, and the little-understood cyclic abundance of the prey species. Adverse conditions of one or more of these factors led to low and even zero chick production. When conditions were favorable, many pairs not only bred successfully but also fledged two chicks. The differences in the feeding regimes of the two species accounted for the small numbers of breeding Brown Skuas as opposed to the large numbers of South Polar Skuas, and for the Brown Skua's uniform productivity between seasons as opposed to the South Polar Skua's fluctuating breeding success.

Pietz (1986) stated that resting and foraging behaviors constituted the two largest time budget components of South Polar Skuas. Although the timing of these behaviors varied greatly from day to day for each individual, there was no hour in which all birds were foraging or at rest. Despite individual variations, both breeding and nonbreeding pairs showed a short peak in resting behaviors around 2400 hours, while foraging activity moved from lowest to highest levels between 0000 and 0500 hours, then gradually declined toward midnight. This pattern held true for males especially, since females often remained on territory during prelaying and early incubation, while their mates procured food at sea for both of them. Foraging-trip length reflected the variable availability of their food resources. Trips away from the territory averaged 7.6 hours during a period of extensive pack-ice cover, whereas foraging trips averaged only 2.4 hours at other times when ice was not a factor. Although female foraging time increased during the later stages of the breeding cycle, only rarely did both parents leave the territory unattended at the same time. Therefore, the timing of an individual's foraging activity was correlated negatively with that of its mate even into the postbrooding period. Nonbreeders and failed breeders correlated more positively.

Other factors that apparently influence South Polar Skua diurnal foraging patterns are near-freezing temperatures and particularly the ability to forage at low light levels—a much debated subject that has produced at least five hypotheses. Pietz (1986) stated that the hypothesis that fits the Palmer study area best is one that relates light intensity and sun position to the visibility of prey below the water surface. As the sun position moves toward the horizon, the portion of light reflected from the water's surface increases and the proportion penetrating the water decreases. Reflection increases when the water surface is disturbed, as during high winds or storms. The comparatively high position of the sun favors the Palmer area (64°-65° S) over skua breeding areas at higher latitudes, such as Ross Island (78° S). The high proportion of cloudy days at Palmer reduces reflection off the water, but conceivably this advantage is diminished by a loss of light intensity for seeing prey.

The coldest temperatures of the day evidently inhibit skua foraging in other regions (Young 1963) due to icing on the foreheads and beaks of the birds, a condition that has not yet proved to be a problem at Palmer. Pietz concluded that light levels ultimately influenced foraging activity of the skuas, although the proximate mechanism may vary. The midnight low in activity at Palmer probably resulted from the effect of light on prey visibility. The early morning low in activity on Ross Island probably resulted from the effect of light

and air temperature. Not yet fully understood is whether an internal clock in the skua's makeup responds to light cues—a hypothesis that merits further study.

Apart from their foraging at sea, the South Polar Skuas had an opportunistic lifestyle that was highly visible. When given the opportunity, they preyed upon almost anything they could manage. When not in the presence of Brown Skuas they readily exploited nesting penguins, as did the pair that set up territory beside a colony of Gentoo Penguins on the west coast of Anvers Island, where Brown Skuas were absent. Judging by the scattered remains of chicks about the skuas' nest, penguins were an important source of food. Only on rare occasions were the South Polar Skuas seen feeding on penguins within Brown Skua territories.

Both species gathered at refuse piles before the closing of the dumps at Palmer Station. Palmer personnel continued to feed one pair of South Polar Skuas that nested on Bonaparte Point while maintaining a feeding territory on a seaward-facing station balcony from which the pair excluded all skuas and gulls. The male of the pair was recognized by its combination of bands: yellow, aluminum—blue, yellow. "Yaby" was known to everyone at Palmer Station and, when he was found dead several years later,

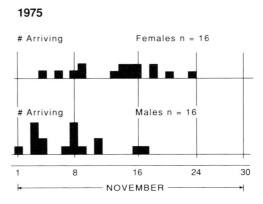

Figure 12. Spring arrival dates of South Polar Skuas (*Catharacta maccormicki*) of known sex at Bonaparte Point, Anvers Island (from Neilson 1983).

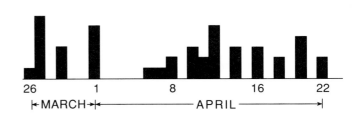

Figure 13. Departure frequency of breeding South Polar Skuas (*Catharacta maccormicki*) from Bonaparte Point, Anvers Island (from Neilson 1983).

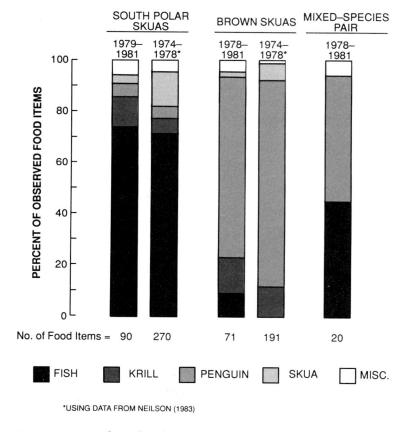

SOUTH POLAR SKUAS

BROWN SKUAS

MIXED–SPECIES PAIR

No. of Food Items = 90 270 71 191 20

■ FISH ■ KRILL ■ PENGUIN ☐ SKUA ☐ MISC.

*USING DATA FROM NEILSON (1983)

Figure 14. Diet of South Polar Skuas (*Catharacta maccormicki*) and Brown Skuas (*C. lonnbergi*) from records of direct feedings, courtship feedings, chick feedings, and regurgitations of food during handling. Total number of food records is given below each bar (from Pietz 1987).

his obituary appeared in the *Antarctic Journal of the United States*. It is noteworthy that the artificial feeding of this pair resulted in much earlier egg-laying dates than for all other pairs nesting close by.

Diet. The primary diet of South Polar Skuas at Palmer was the notothenoid fish *Pleuragramma antarcticum* (Neilson 1983; Pietz 1984) (see Figure 14). Little is known about its movements there in response to light, food, and so on. The notothenoid fishes may follow and feed on krill that move to upper water levels at night and to deeper levels during the day. Such movements are probably less important in areas of nearly continuous daylight, otherwise the skuas at Palmer would have been feeding on krill and fish when they were least available.

Krill, particularly *Euphausia superba*, was an important secondary food source of the South Polar Skuas (Pietz 1984). At times South Polar Skuas and Kelp Gulls congregated near Palmer Station to feed on krill swarms in Arthur Harbor. Large feeding flocks were not encountered far at sea, where single or several widely spaced individuals were apparently the rule. Occasionally the skuas fed on fairy shrimp in freshwater pools. Other prey items recorded in this study included eggs and chicks of

Adélie and Gentoo penguins, Kelp Gulls and Antarctic Terns; highly suspected but not documented were eggs and chicks of Southern Giant Petrels and Antarctic Blue-eyed Shags. South Polar Skuas also forced flying shags to regurgitate their food (Maxson and Bernstein 1982). Their scavenging on seal carcasses appeared to be exceptional.

Cannibalistic behavior and related instances of siblicide were recorded infrequently and only during periods of food stress. Neilson (1983) observed firsthand some cannibalistic attacks by marauding adults, siblings, and chicks from neighboring territories. Remains of older young in the aftermath of cannibalistic episodes littered the territories, but younger chicks simply vanished. During another stress period, Pietz (1987) found the remains of chicks on fourteen of twenty territories under surveillance; some chicks had been eaten on their natal sites, others on neighboring territories. According to Eppley and Rubega (1989), the 1989 oil spill at Arthur Harbor disrupted the normal parental behavior of South Polar Skuas and indirectly caused reproductive failure. They hypothesized that unusual lapses in nest attendance by oiled parent birds resulted in severe cannibalistic losses. However, their hypothesis was challenged by W. Trivelpiece (see Barinaga 1990), who believed

Chinstrap Penguin

Gentoo Penguin

Rockhopper Penguin

Antarctic Blue-eyed Shag

Antarctic Blue-eyed Shag

South Polar Skua (light phase)

South Polar Skua (dark phase)

Hybrid Skua

that food scarcity, not oil, was the real cause of skua reproductive failure that occurred in many places that season. The argument has not been resolved, but evidence from this study indicates that whatever the source, any disrupting influence during nesting results in diminishing reproductive success.

Molt. South Polar Skuas studied by Neilson (1983) showed only a slight body molt prior to migration. Molt of the flight feathers did not occur; of the many individuals handled, only two had replaced their first outer primaries before leaving the area. The delay probably relates to the species' long migrations. Premigration flight-feather molt, such as occurred in the Brown Skuas at Palmer, is thought to be nonadaptive for long-distance migrations (Salomonsen 1976).

Predatory Behavior. South Polar Skuas preyed mostly on fish and krill near the surface of the sea. They fed most often while swimming and dipping their bills, less often while fluttering in midair just above the surface and suddenly dipping. On rare occasions they were seen plunging like terns from heights of 5 meters, becoming partially submerged on impact. Complete submersion was not observed, although at the Falkland Islands DFP several times saw a different species of skua disappear completely beneath the water from a swimming position.

Individual pairs were proficient at nest thievery. During 1974-75 on Bonaparte Point, DFP observed a fairly large Antarctic Tern colony (seventeen nests) that was destroyed completely not by members of a large skua club residing close by, but by a single pair of neighboring South Polar Skuas. Little by little the tern eggs and chicks disappeared, with chick bands later appearing in the skuas' nest. That same season on Humble Island a somewhat smaller tern colony (twelve nests) was gradually wiped out by a single pair of South Polar Skuas that resided close by, based on direct observation and tern bands recovered at the skuas' nest. Although the terns constantly attacked any flying skua, their best defense was to nest far from any skua territory. Pietz (1987) reported that one pair of South Polar Skuas under her surveillance was particularly proficient at catching adult Wilson's Storm-Petrels.

At times the Kelp Gulls flew hard at the skuas and struck them in midair, but in the presence of humans they were prone to leave their nests unguarded, and the skuas took advantage. Consequently, extreme care had to be taken while checking gull nests. South Polar Skuas pursued recently fledged gulls, sometimes successfully forcing them down to the ground or water.

Kleptoparasitism by the South Polar Skuas was described earlier, in the section on the Antarctic Blue-eyed Shag (Chapter 13), which illustrated how this type of behavior was intensified when the skua's usual prey at sea was hard to procure. Since the Brown Skua monopolized the penguin food resource, few observations were obtained on predatory attacks by South Polar Skuas on nesting Adélie Penguins. Nevertheless, all observers in this study concluded that the South Polar Skuas at Palmer were not very adept at procuring penguin chicks.

Breeding Biology

Nest sites. In the Palmer area, South Polar Skuas nested independently of the penguin colonies that were occupied by Brown Skuas. The former occupied territories that ranged from snow-free areas near the sea to inland ridges, including the high cliffs of Hermit Island, where some sites were reminiscent of eagle aeries. Solitary pairs were common enough, but where the species was abundant most

Figure 15. Frequency of five types of long call displays from films of 30 South Polar Skuas (*Catharacta maccormicki*) and 12 Brown Skuas (*C. lonnbergi*), according to Pietz (1985). Percentages represent the number of individuals giving displays of that type divided by the sum of those numbers across types (sums were 45 for South Polar Skuas, 20 for Brown Skuas). The line drawings represent the most extreme backward position of head and body exhibited for each type during the displays.

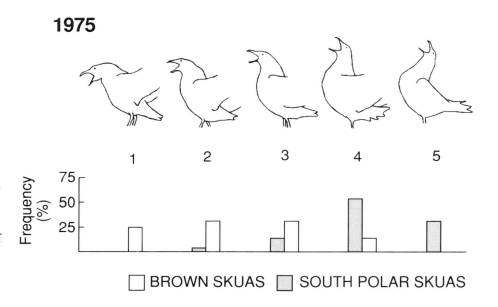

nested in loose to unusually dense colonies for a predator. South Polar Skuas nested on different substrates varying from bare rock to moss carpets, often ranging much farther from penguin colonies than did Brown Skuas.

Density. The densest concentration of South Polar Skuas observed anywhere in Palmer Archipelago occurred within the Palmer study area, presumably because of its high food potential and numerous nesting sites. Although numbers of breeding pairs fluctuated between seasons, certain areas continued to hold the majority, notably Litchfield, Christine, and Shortcut islands and Bonaparte Point. The number of pairs, however, did not necessarily correlate with closeness of nests. On Bonaparte Point, where a number of pairs had been monitored since 1974, some nests were widely separated, and the closest ones were only 12 meters apart. Two of only seven nests recorded for Torgersen Island were a scant 2.5 meters apart, and a third was within 7 meters of those two (Pietz 1987). Despite

the closeness of some of these territories, each was defended tenaciously. Unlike some pairs of Brown Skuas that defended both nest and feeding territories, the South Polar Skuas maintained only nest territories. A notable exception was the pair that was fed artificially at Palmer Station following the closing of its open dumps (see the section above on foraging behavior).

Site and mate fidelity. Subsequent observations of many adults banded during 1974-75 confirmed the belief that the South Polar Skuas generally returned to former nesting territories and mates. Although some pair-bonds were disrupted by the death or long absence of a mate, other pairs mated for several years. Pairs tended to use the same territory each year but often changed nest scrapes within the territory. Changes in territorial dimensions, including locality changes, occurred, but the birds rarely nested far from their former sites. Some pairs that had been banded in 1974-75 still occupied their old sites ten years later.

Defense behavior. With respect to territorial functions, the "long call" was the display most frequently observed in the species' behavioral repertoire. Moynihan (1955, 1962) considered the skua's long call to be homologous with the long call and "oblique" postures of gulls. Beer (1970, 1972) went a step further in stating that gulls also use this display for individual recognition. In this study, the skuas used the long call for greeting mates, and also for advertising the nest territory. A displaying mate invariably triggered the display in its mate. Nearly simultaneous displays of a defending pair occurred whenever their territory was invaded by intruders.

Difference in long-call renditions among the various skuas has stimulated a number of researchers to study these displays (Devillers 1978; Jouventin and Guillotin 1979; Cramp and Simmons 1983). In this study, Pietz (1985) made a detailed comparison of South Polar and Brown skuas with the use of sound and film recordings. Superficially the two species showed very similar visual and vocal long-call displays: the birds typically extended their wings fully backward and usually upward in displaying the white wing bars while vocalizing loudly with bills thrust forward, often tilted either downward or skyward. They sometimes assumed a bent neck posture, quite often with the bill pressed against the breast, both before and after raising the wings, sometimes only before and sometimes only afterward.

In all visual and acoustic details measured, skuas showed much variability within and overlap between the species. Nevertheless, on average, the two species differed in several respects. With respect to postural aspects of the long call, only the most extreme postures appeared to be species specific (Figure 15). South Polar Skuas tilted their heads noticeably farther back and more vertically than did any of the Brown Skuas observed by Pietz (1985). Conversely, Brown Skuas often assumed a far forward position rarely seen in South Polar Skuas.

Calling postures sometimes varied depending on the location of an intruder or a mate. Pietz (1985) noted that South Polar Skuas leaned slightly forward when facing their mates, and leaned further back when calling beside and parallel to their mates. Facing an intruder on the ground, the defending South Polar Skua leaned or walked toward its adversary, calling as it went.

With respect to acoustic aspects of the long call, Pietz noted that South Polar Skuas, on average, produced notes with a faster repetition rate, lower pitch, lower average frequency, and more harmonics than those of Brown Skuas. Until more skua populations are studied, the evolutionary significance of these minor differences in display postures and call notes cannot be

fully assessed. The differences are slight enough to be of marginal value for field identification of the species, though some of the long-call components may be useful for individual recognition.

Although the South Polar Skuas long-called from almost anywhere near their eggs or chicks, they typically performed from what appeared to be favored perches, usually elevated spots. If they failed to thwart the advance of an intruder through long-calling, they quickly left their perches and gave chase, frequently following a fleeing skua or gull far from the territory. Human intrusion almost invariably resulted in the dive-bombing pursuits for which skuas are notorious. About 50 percent of the dive-bombing skuas were easily caught in hand nets thrust above one's head. Another 40 percent required time and patience to catch. The remaining 10 percent proved uncatchable.

Agonistic displays and encounters, according to Pietz's (1986) time-budget studies, occurred during nearly all hours. A lull in these activities during twilight was probably due to a corresponding lack of intruders.

Sexual displays. Other than the mate greeting function ascribed to the long-call display cited above, Pietz (1984) observed that during the pre-egg-laying period, females received most of their food through courtship feedings, which probably were prerequisite to successful copulation. She also observed pairing behaviors sometimes referred to as the "squeaking ceremony," though she rarely heard the squeakings.

Nest-building. Nests ranged from unlined scrapes in the turf or bare soil to rather elaborate bowl-shaped structures heaped with plucked chunks of moss interspersed with lichens and grasses. The tossing of moss or other material sideways by either sex contributed to nest construction, but, according to PJP, it also functioned in pairing and occurred as displacement behavior following a disturbance.

Egg-laying. Clutch initiation dates varied between seasons due largely to food availability. Over the six seasons observed by Neilson (1983) and Pietz (1987), the South Polar Skuas exhibited a four-week spread in mean initiation dates. Delayed egg-laying, as was the case in 1979-80, coincided with limited access to food, usually because of ice cover. Conversely, the earliest laying dates were achieved by the pair of South Polar Skuas that received an artificially supplemented diet (food scraps) from Palmer Station. Their chicks hatched a week before the next earliest pair and about three weeks ahead of the median date for

the species. Early laying resulting from artificial feeding also occurred in the South Shetland Islands (Trivelpiece et al. 1980).

Asynchronous layings within and between seasons at Palmer resulted in a large spread in the occurrence of South Polar Skua eggs. The earliest date recorded for an egg was 28 November 1975, although back dating from hatching indicated that laying started sometime during mid-November in 1976 (Neilson 1983). The latest date recorded for a viable egg was 16 February 1983. South Polar Skua pairs that lost their eggs usually did not replace them, although they continued to defend their territories for some time. The one known exception occurred at Humble Island in 1984-85, when, following the loss of her eggs during 28 December-4 January, the female of a banded pair mated with a different, unbanded male near her original nest and produced a second clutch of two eggs that were fresh when first seen on 19 January. It was not determined whether the repeat nesting succeeded (DFP).

The interval between layings in two-egg clutches was not determined in this study, but, judging by sibling hatchings, it was two to three days.

Clutch size. Food stress resulted in a small second egg of the usual two-egg clutch, or in no second egg at all. Only 45

percent of the South Polar Skuas in Pietz's (1987) sample produced two-egg clutches during an early food shortage in 1979-80, and a significant number of them laid small second eggs. All the pairs of her sample in 1980-81 laid two-egg clutches, with no significant differences between sibling eggs when food was sufficient.

Incubation. During pre-egg-laying and early incubation, females tended to remain at the nest while the males were off procuring food, but thereafter males attended the eggs more frequently. Overall, males were responsible for 46 percent and females for 56 percent of 271 hours of incubation observed by Pietz (1984). Rarely were both sexes away from the nest at the same time. Pietz (1986) observed that simultaneous absences occurred for less than 3.5 minutes of the 284 hours of observations on six pairs.

The precise incubation period from laying to hatching of the last egg was not determined in this study, but approximations for three clutches ranged from 28 to 33 days. According to Watson (1975), the period is 24 to 34 days.

Hatching. Known hatching intervals for three nests ranged from 52 hours to 67 hours, indicating that incubation started soon after the laying of the first egg. During four seasons (1974-78), DRN recorded hatching dates from late December to late January. Pietz (1987) found the range to be 13 January-9 February (mean 27 January) at fifteen nests in 1980, and 3-25 January (mean 14 January) at nineteen nests in 1981. DFP and CCR recorded hatching dates from 29 December 1983 to 11 February 1984, further demonstrating the wide range of hatching dates in the Palmer area.

Brooding. Females brooded somewhat more than their mates. Of 124 hours of brooding behavior recorded by Pietz (1984), 59 percent was by females. Chick survival depended on unwavering vigilance of the parents. Pietz (1986) noted that successful pairs rarely left the chicks unattended (less than 5 minutes during 112 hours of observation).

Fledging. Several marked chicks in this study provided approximate time intervals from hatching to strong flight. A chick hatched on 8 January 1976 fledged by 7 March (58 days); one hatched on 9 January 1977 fledged by 5 March (55 days); one hatched on 10 January 1977 fledged by 10 March (51 days); one hatched on 13 January 1980 fledged by 10 March (57 days). These intervals fall within the range of 49 to 59 days stated by Watson (1975).

The earliest date recorded for fledging was 12 February 1976. A conservative fledging period of only 50 days indicated that some late-hatching chicks probably did not fledge before early April—at a time when juveniles from early nestings were already migrating.

Postfledging. Neilson (1983) observed that siblings migrated together from their natal sites soon after they had gained independence from their parents. He witnessed the 1975 exodus of 37 fledglings from Bonaparte Point: nearly 60 percent from early nestings left during 12-13 March, 32 percent from later nestings left during 2-10 April, and a few left during 25-28 April. An individual from an early nesting that hatched on 28 December and fledged 55 days later on 21 February would have been 74 days old when it left its natal ground on 12 March.

Breeding success. No species in this study showed greater between-year reproductive variability correlated with variability in food supply. As a result, the South Polar Skuas showed wide ranges in nest density, percentage of two-egg clutches, and number of fledglings/pair. When foraging conditions were optimal throughout the season, large numbers mated and produced two offspring. Poor to marginal conditions early in the season resulted in fewer nesting

attempts and a reduction in egg size and clutch size. Poor to marginal conditions following the hatch resulted in longer foraging trips, reduced nest attentiveness, chick starvation, and sibling conflict, and were invariably followed by reduced numbers of fledgings. So far as is known, not one fledgling was produced during the 1977-78 or 1982-83 seasons.

Synopsis of annual cycle. Commencing in late October, South Polar Skuas returned to the breeding grounds over a period of several weeks. Males, on average, arrived ahead of their mates at the nest sites, which, when Brown Skuas were present, were not associated with penguin colonies. Site fidelity and mate fidelity were pronounced. Both sexes defended their nest sites through oft-repeated long-call displays and dive-bombing tactics, especially while incubating eggs and caring for the chicks, which were rarely left unattended. Reproductive success was tied closely to the availability of fish and krill obtained at sea, resulting in extreme highs and lows in the number of fledglings produced in different seasons. The breeding cycle from clutch initiation to fledging was extended from egg-laying as early as mid-November to fledging as late as early April—a range of 140 or more days. Fledglings left their natal

Brown Skua

Brown Skua
(*Catharacta lonnbergi*)

Overall Breeding Distribution. Circumpolar. Breeds chiefly in the Sub-Antarctic, but also along the Antarctic Peninsula to about 65° S, and possibly even south of the Antarctic Circle at the Balleny Islands.

Current Status in Palmer Archipelago. Brown Skuas were common, conspicuous birds in the South Shetland Islands, but their breeding distribution appeared to be spotty in Palmer Archipelago. The species was well established along the south coast of Anvers Island from Biscoe Point to and including the Joubin Islands. Mixed matings with South Polar Skuas produced some hybrids, although generally the two species remained separated. Brown Skuas migrated from the region, but their whereabouts in winter are not known. Most of the observations in this study were by Pietz and Maxson (1980), Pietz (1982, 1984, 1985, 1986, 1987), Neilson (1983), Parmelee (1985), and Parmelee and Pietz (1987).

Banding. During 1974-85, 158 Brown Skuas were banded within the Palmer study area, and an additional 30 were banded elsewhere along the south coast of Anvers Island (Table 20).

grounds in waves over a period of six or seven weeks, starting in mid-March. They migrated separately from the adults, and wintered as far north as Oregon in the North Pacific and Greenland in the North Atlantic. Banded individuals of both sexes later returned to their natal areas as breeding adults when five years of age or older. Paired individuals departed separately from their breeding grounds, females ahead of their mates, and migrated to wintering areas yet to be determined.

Adults. Of the 72 adults banded, 47 were already paired at nest sites near Adélie Penguin colonies. There were 21 single individuals banded at Palmer Station, Old Palmer, and Torgersen Island; 9 of them were later sighted at nest sites either within or outside of the study area. Four additional adults were paired with South Polar Skuas. DRN color banded and sexed many of these birds during the early years of this study. To date none of the 72 adults has been recovered beyond Anvers Island.

Nestlings. During 1974-85, 116 of the 126 nestlings banded were believed to have fledged. To date none of the 116 has been recovered beyond Anvers Island. A few returned to the study area as prebreeders; only 4 are known to have returned as breeding adults:

A. DFP and CCR recovered female 877-36550 (rebanded 1067-25097) at its nest on Humble Island on 3 January 1984 within 1 kilometer of its natal nest on Litchfield Island, where it had been banded by DRN eight years previously, on 18 January 1976.

B. DFP and CCR recovered female 877-36934 at its nest in the Joubin Islands on 12 January 1984 about 15 kilometers from its natal nest on Christine Island, where it had been banded by DRN seven years previously, on 12 January 1977.

C. DFP and JMP recovered female 1067-25067 at Dream Island on 12 December 1984 at its natal nest scrape, where it had been banded by DFP six years previously, on 5 February 1979.

D. D. G. Ainley (personal communication) observed female 1067-25055 at its nest on Litchfield Island on 2 January 1988 within 4 kilometers of its natal nest on Christine Island, where it had been banded by DFP eight years previously, on 18 February 1980.

Summer Records. Within the Palmer study area, Brown Skua pairs numbered from 6 to 12 per season; an average of 8.6 per season produced eggs. Most of these birds were concentrated within the triangle formed by Torgersen, Litchfield, and Humble islands, a fairly small area but one that contained 84 percent of the area's penguin nests (see Figure 16). One or two pairs usually nested on Christine Island, and occasionally one pair on Cormorant Island. Elsewhere along the south coast of Anvers Island, at least two pairs bred at Biscoe Point and two pairs at Dream Island. Although only one nest was located in the Joubin Islands, several encounters with Brown Skuas at various times suggested that more than the one pair bred there. The species was scarce along the west coast of Anvers Island, where DFP noted a single, nonnesting pair on territory

Table 20. Summary of Brown Skua (*Catharacta lonnbergi*) bandings along the south coast of Anvers Island during 1974-81 and 1984-85

	1974-75	1975-76	1976-77	1977-78	1978-79	1979-80	1980-81	1983-84	1984-85	Total
Adults										
Litchfield Island	13	2	1	2		2		6	1	27
Palmer Station	17									17
Humble Island	2	2	1					2		7
Cormorant Island	3			1						4
Biscoe Point									4	4
Christine Island			1				1		1	3
Torgersen Island						3				3
Dream Island								3		3
Hermit Island	1									1
Norsel Point						1				1
Old Palmer						1				1
Joubin Island								1		1
Total	36	4	3	3		7	1	12	6	72
Chicks										
Litchfield Island	10	11	10	9	8	7	2	10	1	68
Humble Island	2	1	2	2	2	3	2	2	2	18
Dream Island			8	3				4	3	18
Christine Island			1		2	3		3	2	11
Cormorant Island	2	1	1			2				6
Biscoe Point									5	5
Total	14	13	14	19	15	15	4	19	13	126

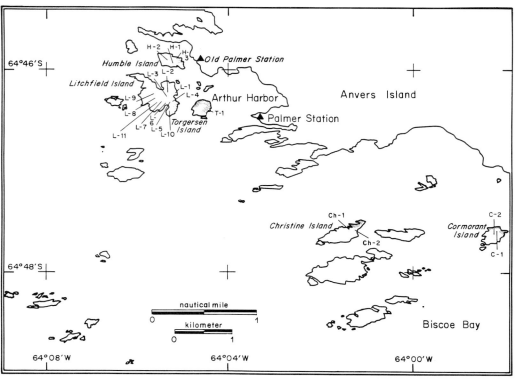

Figure 16. Palmer study area with approximate nest sites of Brown Skuas shown on Cormorant, Christine, Torgersen, Litchfield, and Humble islands. Adélie Penguin colonies on these five islands are indicated by areas shaded with parallel lines.

near Gerlache Point on 2 January 1985. Beyond Anvers Island records were decidedly scarce. Parmelee and Rimmer (1985) noted a nest with one egg near Astrolabe Needle, Brabant Island, on 29 December 1983, and DFP noted a nonnesting pair on territory at D'Hainaut Island, Mikkelsen Harbor, Trinity Island, on 16 January 1989.

In this study individuals were sighted as far south as the Argentine Islands, but the only definite nesting recorded close to the Antarctic Peninsula was at Cuverville Island (64°41' S, 62°38' W), where DFP noted one large downy chick on 26 January 1989.

Arrival. Earliest return recorded for the Palmer study area was 25 October (1975). Neilson (1983) determined that the average first date for eight males and seven females was 7 November (range, 30 October-12 November); in six observed cases, four males and two females arrived at their nesting territories ahead of their mates.

Departure. Neilson monitored the departure of Brown Skua pairs at eight nesting sites in 1975 and showed that paired birds do not necessarily leave together. Although one bird may have left as early as 6 March, the majority departed their nesting territories between 16 March and 8 April. In three cases, paired birds left about the same time, but whether they left still paired is not known. Individuals from the remaining five pairs left separately, in some cases several days apart. Not determined was whether their fledglings went with them. The fact that some adults arrive and depart separately from their mates suggests that the pair-bond functions only on the breeding ground.

Territories. Brown Skua pairs defended the area surrounding the nest, and the majority of them also defended feeding territories (penguin colonies) nearby. Several pairs without feeding territories on penguin-poor Litchfield Island regularly flew to Torgersen Island, where there were many penguins but no discernible skua feeding territories. Not completely resolved was whether the Brown Skuas with feeding territories adjacent to nest sites had an advantage over those that lacked them. In studying the advantages of territories with or without penguin colonies, Neilson (1983) observed that even though foraging time was greater for the latter, the young from each category developed at about the same rate. The pairs without feeding territories actually fledged the most young—an observation contrary to the findings of Trillmich (1978) and Trivelpiece et al. (1980).

Diet and Foraging Tactics. Both male and female Brown Skuas foraged in penguin colonies, sometimes cooperating to distract an adult penguin from its egg or chick. They also cooperated in tearing carcasses to pieces with their beaks while rarely using their feet in manipulating prey. Neilson (1983) and Pietz (1986) observed that pairs with feeding territories adjacent to their nests foraged less (8-13 minutes per bout) than those without feeding territories (16-60 minutes). Through most of the season they spent less time foraging than did the South Polar Skuas. Following the fledging of penguins, their foraging time increased significantly in order to satisfy the demands of their developing chicks. At this stage, Brown Skua parents spent more time scavenging and utilized a greater variety of prey species. Both skua species foraged in a melt pond during a temporary abundance of fairy shrimp (*Branchinecta* sp.), but Brown Skuas generally fed on krill spillage in penguin colonies rather than at sea. A few Brown Skuas may even have fished at sea,

for Pietz observed a single female feeding five fish to its chick in one 24-hour period. Evidently the Brown Skuas of the South Shetland Islands utilized more fish than did those at Palmer (see Peter et al. 1990).

In a mixed-species pair, Pietz (1986) observed that the Brown and South Polar skuas showed foraging behavior characteristic of their respective species. The Brown Skua female fed primarily on neighboring islands, and averaged trips of less than 1 hour while penguins were present, but more than 3 hours following their departure. Its South Polar Skua mate foraged at sea and averaged 2.8 hours per trip when there was little ice, but considerably more during periods of extensive ice cover. Both members of this pair fed the hybrid chicks. Over two seasons, food records from the nest site (e.g., observed chick feedings) showed their food to consist of 50 percent penguin and 45 percent fish. According to Peter et al. (1990), fish was the main food of mixed pairs in the South Shetlands.

With respect to penguin kills, Brown Skuas usually carried stolen eggs some distance from the colony before eating them, often at their nest sites. Downy chicks were swallowed whole usually on the spot. Large chicks were dragged a short distance, killed, and eaten over several hours. According to

Pietz, the skuas left only feet, bones of the legs and pelvic girdles, and inverted pelts.

Apart from the penguin colonies, seabirds were a source of food for a pair of Brown Skuas nesting on a sea cliff near Astrolabe Needle, Brabant Island (Parmelee and Rimmer 1985).

Molt. Neilson (1983) found a significant difference in the timing of the molt in the two species of skuas. Although both underwent a limited body molt prior to departure, Brown Skuas also had a fairly extensive molt of certain flight feathers. In Brown Skuas, molting began in early January in the abdominal region and gradually extended onto the throat. Back feathers were replaced next; by mid-February molting appeared on the nape and head. Molt of flight feathers was first detected in early February, when several individuals showed incoming first (innermost) primary feathers. Second and third primary feathers were also molted on the breeding grounds. The loss of these feathers resulted in a conspicuous gap in the wings up to the time of departure, although one individual already had a fully developed first primary feather when last seen on 4 April. The fact that South Polar Skuas did not show a similar loss of flight feathers suggests that the Brown Skuas have a much shorter migration than the other species.

Mortality. Of 50 nesting Brown Skuas banded at Palmer, 30 percent were known to have died over the years from disease or other causes. Another 40 percent failed to return, presumably having died while away, unless absent from the breeding grounds longer than expected. For reasons unknown, twice as many males as females disappeared or were found dead. The remaining 30 percent of the breeders survived through 1984-85, when they were last checked. Of the 116 chicks believed to have fledged, only 4.3 percent had returned up to that time. To date there have been no recoveries at other skua nesting areas beyond Anvers Island, indicating high postfledging mortality.

During 4-8 February 1979, as many as 22 Brown Skua adults may have perished in the study area from a virulent strain of fowl cholera (*Pasteurella multocida*) (Parmelee et al. 1979). Half of these birds appeared to be nonbreeders; the rest were known breeders, including 9 banded individuals with documented breeding histories. Fowl cholera is characterized by inverted age resistance, and the chicks outlived their parents as expected. Dependent young that lost both parents probably died when left unattended.

As a result of the fowl cholera epidemic of 1979, and a disturbance from a parasit-ological collecting expedition in 1983, several traditional Brown Skua nesting territories were left vacant. Reoccupancy following these sudden vacancies proved to be slow and indicated that the integrity of skua boundaries may persist for one or more seasons even when unoccupied. The vacated sites were reoccupied by surviving and new Brown Skuas, and in some cases by South Polar Skuas. Several territories remained unoccupied for at least two seasons. One surviving female with a new mate occupied her old site the season following the cholera epidemic, but the pair failed to produce eggs until a year later.

Breeding Biology.

Nest sites. Nests were fairly close to bird colonies, chiefly penguins. They occurred from low-lying areas near the beaches to rocky prominences and ridges, even formidable cliffs. Most observed nests were mere scrapes, sometimes in barren areas, but more often in luxuriant vegetation that extended several meters out from the scrape, creating a conspicuous green patch in an otherwise austere surrounding. Brown Skua sites were marked by the scattered remains of eggs and chicks of their prey species.

Density. The densest numbers occurred on eastern Litchfield Island, where as many as seven pairs bred one season (1977). Brown Skuas generally nested well apart, and even during the crowded seasons at Litchfield Island, the nearest neighbors were separated by 30 meters or more.

Site and mate fidelity. With few exceptions Brown Skuas remained faithful to their nest sites. Of the many adults banded at their nests, only two males and three females changed territories: three simply to new areas close by on the same island, and two to adjacent islands. Continuous residency by the same pair was five years at three sites when terminated by fowl cholera. The longest recorded residency at one site by an individual (female) was eleven seasons, assuming that the bird was present during two seasons not checked.

Mates generally remained faithful, but there were exceptions. Each of the five recorded changes in territory was accompanied by a change of mate. In twenty-one other cases, birds changed mates without changing territories. It is not clear why such individuals changed territories and mates. Different arrival times of the sexes from one season to the next may have had a disruptive influence, but mate and site changes more likely were prompted by failure of a mate to return to the territory. In many cases birds that did not return are known to have died; in other cases their fate was unknown. In some cases, former breeders returned to the area but did not nest, or they were not seen in the area for one or more seasons before returning to breed.

Trios. The occupancy of trios (three adults) at a nest site was rare in this study. Of the four recorded instances at Palmer, there was but one fledgling resulting in nine seasons, indicating that trios were not nearly so common and productive in this region as they have been elsewhere (see Young 1978).

Defense behavior. Brown Skuas defended nesting and feeding territories tenaciously against all intruders, including humans. The long call was the most common, audible, and visible display; it functioned both as a mate greeting and as a territorial advertisement. One or both members of the pair stretched the wings upward and backward while calling loudly, often from some promontory within the territory. Brown Skua long calls closely resembled those of South Polar Skuas, with subtle differences, as discussed in the section on the latter species.

High-energy agonistic encounters took the form of swift aerial chases and fights either in the air or on the ground. Pietz (1987) observed that these chases were significantly more frequent for Brown Skuas with feeding territories. Agonistic encoun-ters ranged from 0.80 per hour to 1.75 per hour for those with feeding territories, and from 0.08 per hour to 0.33 per hour for those lacking them. According to Pietz (1986), there was no significant correlation between level of agonistic activity and time of day. Although agonistic displays and encounters occurred during nearly all hours, there was a lull at twilight, when intrusion was down. Where South Polar Skuas nested close to Brown Skuas, in some cases only 15 meters apart, aggressive inter-action between the two species was nearly continuous. Litchfield Island, with its many nesting skuas, was a hotly contested ground; DRN witnessed the killing there of several South Polar Skuas by a particularly aggressive pair of Brown Skuas.

Nest-building. It was not determined in this study whether one or both sexes par-ticipated in nest construction. Brown Skua nests were often mere scrapes in the turf without much added material. Traditional sites often had several scrapes used in vari-ous years. Newer nests, particularly those in moss carpets, were made by pulling out chunks of the vegetation. The loose chunks deposited around the rim of the bowl pre-sented a well-defined nest.

Egg-laying. Brown Skua clutch initia-tions were generally linked with those of their prey species. In the Palmer study area, the Brown Skuas got off to an early start in concert with the Adélie Penguins. The ear-liest date recorded was 29 November 1977, although back dating from measured chicks in 1976 indicated that eggs had been laid as early as 19 November. Neilson (1983) determined that the within-season range in egg-laying was 8-19 days for three seasons, the mean dates falling between 30 Novem-ber and 5 December. An unusually late nesting extended the range of clutch-initiation dates to 41 days in Pietz's (1987) study. Nevertheless, Neilson's mean clutch-initiation dates fell within 4 days of those recorded by Pietz, resulting in a five-season range of only one week.

Taking into account the several late nestings, the range in egg-laying was con-siderable, probably from as early as 19 November to as late as 8 January. In the South Shetland Islands during 1983-85, Peter et al. (1990) recorded the range in egg-laying from 23 November to 30 Decem-ber. Conceivably, the late nestings recorded at Palmer by Pietz (1986) resulted from the disruptive influences caused by the fowl cholera epidemic of 1979. The later col-lecting of Brown Skuas for parasite studies almost certainly had an impact upon the Palmer population as well. Another possibility is that the late clutches were replacements from earlier losses, though renesting did not occur among the other Brown Skuas known to have lost clutches in this study. Nevertheless, two late nestings observed by DFP on Litchfield Island in 1984-85 were suggestive. When first observed on 11 December, both pairs defended traditional, but eggless, sites at a time when others had well-incubated eggs. The pair at Site L-10 had nested there the previous season and, close to their original scrape, finally produced a single egg (laid about 28 December) that hatched on 27 January. L-7 had a previous occupant and a new, unbanded one that finally produced two eggs about 50 meters from the old site; the first of these also hatched on 27 January. The situation at L-7 was reminiscent of the documented case of renesting at a South Polar Skua's nest following the replacement of a mate within the season.

Clutch size. Mostly two-egg clutches were recorded in this study, along with occasional single-egg clutches. No three-egg clutches were recorded. Where the clutch was believed to be complete at seven nests in 1979-80, and at six nests in 1980-81, there was one single-egg clutch each season (Pietz 1986).

Incubation. Approximate incubation periods were recorded for two eggs: an egg laid on 30 November was pipped on 30 December in 1977—a minimum of 30 days; another laid on 29 November was hatched by 29 December 1980—a period no greater than 30 days. Females incubated the most at several nests observed by Pietz (1984); during 132 hours of observation, two females incubated 57 percent of the time.

Hatching. Although the range in hatching was considerable (24 December to 7 February), most eggs hatched in late December and early January. Pietz's (1987) median hatching dates for two seasons were 27 and 30 December. Sibling eggs at one nest hatched at 0955 hours on 6 January and at 0921 hours on 8 January in 1978—about two days apart. The hatch interval was roughly two days at two other nests in 1980.

Brooding. Both sexes brooded. Although it was not determined which one brooded the most, the activities of the male and female were coordinated so that one parent was present while the other was away.

Fledging. Early fledging dates for strong-flying young recorded in the Palmer study area were 22 February 1975 (one bird), 24 February 1977 (four), 25 February 1977 (three), 28 February 1977 (three), 21 February 1980 (one), and 16 February 1984 (two). These dates fall close to those

recorded for the species in the South Shetland Islands by Peter et al. (1990), namely 23 February 1984 and 20 February 1985.

An unfledged young was recorded at Palmer as late as 30 March from an egg that had hatched on 7 February 1980. Although the period from hatching to fledging was not determined precisely, it was no more than 54 and 55 days for a pair of siblings in 1979-80, and approximately 52 days for the older of two siblings at another site in 1983-84. DFP noted that strong-flying fledglings remained faithful to their natal sites for 11 or more days. Although they circled widely, they always returned within a short time. The adults continued to defend them, but it was not determined how long the adults fed them.

Breeding success. Brown Skua productivity did not vary much from year to year, as clearly indicated by the banding record of fledglings. The penguin-dependent birds, with their highly stable food supply, did not encounter the difficulties faced by the sea-foraging South Polar Skuas. The one exception was the 1979 fowl cholera epidemic, which devastated the Brown Skuas.

Of 74 Brown Skua nestings (1974 to 1985) in which eggs were found, 71 nests produced 96 fledglings, or 1.35 fledglings per nest; 3 nests produced chicks whose fates were unknown. The fledging of two siblings at a number of nestings indicated an ample food supply accompanied by low siblicide. Neilson (1983) believed that disruptive incubation for whatever reasons resulted in egg mortality and was a major factor in determining the breeding success of Brown Skuas, a conclusion also reached by Burton (1968) at the South Orkney Islands. Egg mortality proved greater than chick mortality in three out of four seasons during Neilson's study.

Synopsis of annual cycle. Brown Skuas usually returned from unknown wintering grounds in early November. Individuals of both sexes arrived at about the same time to traditional nest sites situated near penguin or other bird colonies. The earliest recorded age for breeding was six years. Pair-bonds generally were long-lasting, but some were disrupted through the death or temporary absence of a mate. Because breeding in synchrony with the prey species is thought to benefit Brown Skuas, the occurrence of late nestings suggested disruptive influences, such as diseases and human disturbances. Replacement in vacated skua territories was slow, perhaps because of small numbers and high site tenacity. Breeding success was generally high, and siblicide was rare. The stable annual production of fledglings likely was related to their use of penguins as a primary food source; penguins were con-sistently available and accessible, unlike the fish and krill sought at sea by South Polar Skuas. The Brown Skua's fledging period extended beyond the departures of penguins and forced the adults to find other prey late in the season—a situation that probably favored early fledging. The period from the first estimated egg-laying to the last estimated fledging spanned 127 days; by eliminating a few exceptionally late nestings, it was no more than 112 days, considerably less than the range of 144 days recorded for South Polar Skuas. Brown Skua adults showed fairly extensive molt of flight feathers before departure. The members of some pairs departed separately, but it was not determined whether the young accompanied adults in their departures and migrations. All Brown Skuas were gone before mid-April.

Hybrid Skua
(*Catharacta maccormicki* x *lonnbergi*)

Overall Breeding Distribution. Breeds in the Maritime Antarctic, and possibly in the Balleny Islands.

Current Status in Palmer Archipelago. Small numbers of hybrid skuas occur wherever the breeding grounds of South Polar and Brown skuas overlap. The two species are allopatric over much of their vast breeding ranges in the Southern Ocean, but a few pairs nest side by side and hybridize to some extent within a narrow zone of overlap in the Antarctic Peninsula region (Watson 1975), the South Shetland Islands (Trivelpiece and Volkman 1982; Peter et al. 1990), and, more recently, the South Orkney Islands (Hemmings 1984). In this study, mixed pairs were recorded in the Palmer study area and in other places along the south coast of Anvers Island. These pairs and their hybrid offspring were studied mostly by Neilson (1983), Pietz (1985, 1987), and Parmelee (1985, 1988b).

Banding. Twenty-four F_1 hybrids believed to have fledged within the Palmer study area (Figure 17) were banded as nestlings at the following locations: six at one site on Cormorant Island from 1974 to 1977, twelve at four sites on Humble Island from 1974 to 1985, four at one site on Hermit Island from 1974 to 1976, one on Shortcut Island during 1975-76, one on Norsel Point during 1979-80. Six additional hybrids were banded west of the study area: two in the Joubin Islands at one site during 1976-77, four at one site on Dream Island from 1983 to 1985. Each hybrid was banded with a numbered band on the right leg and a white plastic band on the left leg.

Figure 17. Palmer study area with areas shaded by parallel lines showing locations of Adélie Penguin colonies on Cormorant, Christine, Torgersen, Litchfield, and Humble islands. Mixed pairs of South Polar and Brown skuas produced 24 hybrids on Cormorant Island (Site A), Hermit Island (Site A), Shortcut Island (Site A), Humble Island (Sites A, B, C, D) and Norsel Point (Site A). On Christine Island (Site A), an eight-year-old F_1 hybrid mated with a South Polar Skua and produced an F_2 hybrid fledgling; by age fourteen, it had moved to Site C, where it mated with another South Polar Skua. A mixed pair occupied Site B on Christine Island but produced no progeny during two consecutive seasons.

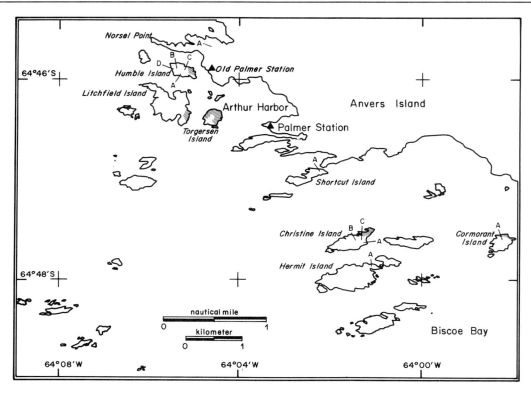

Of the thirty hybrid nestlings banded, there has been but one recovery to date: hybrid 877-36556 (rebanded 1067-25309), banded by DRN on Humble Island on 18 January 1976, was recovered when four years old and set free from a fish net by E. Cunico at Parangual, Brazil, on 23 May 1980. The hybrid (sex uncertain) was observed next by DFP on 2 January 1984, when it was eight years old and mated with a South Polar Skua (banded 1067-25708) at Christine Island; the pair produced one F_2 hybrid, establishing beyond doubt the fertility of *maccormicki-lonnbergi* hybrids. CCR banded the nearly fledged F_2 hybrid 1067-25841 on 24 February. The mixed pair occupied the same site on Christine Island in 1984-85 but produced no progeny; shell fragments at their nest indicated earlier loss of eggs. No additional observations were reported until 10 January 1990, when DFP recorded the then fourteen-year-old F_1 hybrid at a new nest site several hundred meters from its former one on Christine Island. By that time it had mated with another (unbanded) South Polar Skua. According to WRF, the mixed pair's one F_2 hybrid did not survive, along with many South Polar Skua chicks that season. At age fifteen the hybrid returned to Christine Island in 1990-91, but it produced no young that season (WRF).

Summer Records. Watson and Angle (1966) observed a mixed pair at Anvers Island as early as 1966, and in this study they were recorded in the Palmer area every season since 1974. Elsewhere, Neilson (1983) recorded a mixed pair at the Joubin Islands on 7 February 1977. Parmelee (1988b) recorded the following mixed pairs on or near Anvers Island: one pair on Dream Island on 7 January 1984 and again on 5 January 1985, one pair at a nameless island 1 kilometer southwest of Dream Island on 31 December 1984, and one pair at Biscoe Point on 17 December 1984. Hybrids were not produced each year, and none was observed during winter.

Morphology and Growth Rates of F_1 Hybrids. Body size, bill, and tarsal measurements of first generation (F_1) hybrid skuas were intermediate between *maccormicki* and *lonnbergi*, rendering the birds extremely difficult to identify. Most field identification puzzles in this study were attributed to unmarked hybrids or their offspring. Breeding hybrid 1067-25309 had the appearance of a small *lonnbergi*. At the South Orkney Islands, Hemmings (1984) found that hybrid egg sizes and chick growth rates fell between those of *maccormicki* and *lonnbergi,* a conclusion also reached by Peter et al. (1990), who examined hybrids in the South Shetlands. Neither author stated whether the females of mixed pairs were known to be *maccormicki* or *lonnbergi*—an important point considering the large size of *lonnbergi* females. In two breeding seasons at Palmer, PJP also found hybrid growth rates to be intermediate, but for a mixed pair with a known *lonnbergi* female, she found the egg size to be similar to that of *lonnbergi*. Given the small number of hybrid eggs in their samples, this discrepancy probably reflects individual differences in size (and species) of females in mixed pairs.

Taxonomy. The Palmer study area appeared to be fairly typical of Short's (1969) description of a zone of overlap and hybridization where secondary intergradation has occurred: parental phenotypes far outnumbered hybrids, and mixed matings were comparatively rare. Asynchronous arrival and breeding times were apparent premating isolating mechanisms. Although there appeared to be no postmating isolating mechanisms with respect to hybrid fertility, the expected number of mixed pairs through chance mating was calculated to be far greater than the one to four pairs recorded annually in this study. The two skuas show a very close though separate relationship, and thus are considered separate species—a conclusion also reached by Peter et al. (1990).

Breeding Biology of Mixed Pairs.

Nest sites. Mixed pairs followed the pattern of South Polar Skuas in that their nest sites were not necessarily close to penguin or other seabird colonies as was invariably the case with Brown Skuas.

Mate selection. In twenty-three of twenty-five mixed pairings observed in this study, a male South Polar Skua bred with a female Brown Skua. Trivelpiece and Volkman (1982) and Hemmings (1984) found only male *maccormicki* mated to female *lonnbergi*.

Pair and nest site fidelity. Some mixed pairs remained faithful to one another at an established site for at least four consecutive seasons. Either member of a pair might reoccupy a site that was abandoned for one or more seasons and then mate with a new individual of the same or the other species. It was not determined whether a mixed pair ever reunited following absenteeism.

Breeding success. Breeding chronology and breeding success in mixed pairs paralleled that of South Polar Skuas, indicating the important role of the male (usually a fish-dependent South Polar Skua) in securing food for the female and later for the young.

Kelp Gull

Of the more than forty species of gulls in the world, the Kelp Gull (*Larus dominicanus*), also called Dominican Gull and Southern Black-backed Gull, is the only gull that breeds in the Antarctic and Sub-Antarctic. According to Watson (1975), the species is a widely distributed southern representative of the Northern Hemisphere black-backed/ herring gull complex. It breeds abundantly in South Africa, southern Australia, and the New Zealand region, and in South America as far north as 7° S, including the Falkland Islands. Within the western sector it was particularly abundant and conspicuous in many areas throughout the South Shetland Islands, Palmer Archipelago, and in places along the Antarctic Peninsula to 68° S. Adults and their young resided throughout the year in Palmer Archipelago, although some young and possibly some adults migrated as far north as Argentina and Chile.

Watson (1975) lists several vagrant gulls for Antarctica: the Band-tailed Gull (*Larus belcheri*), a South American species recorded on South Georgia in the South Atlantic; the Franklin's Gull (*Larus pipixcan*), a North American species recorded on Gough Island in the South Atlantic; and the Silver Gull (*Larus novaehollandiae*), a South Africa-Australia-New Zealand species recorded on Marion Island in the Indian Ocean.

Kelp Gull
(*Larus dominicanus*)

Overall Breeding Distribution. Circumpolar in Southern Hemisphere. Subspecies *L. d. inicanus* breeds southward in the American sector from southern Brazil and Peru to about 68° on the Antarctic Peninsula; in the Indian sector on Prince Edward, Marion, Cozet, Kerguelen, and Heard islands; and in the Australian sector in Aus-

Figure 18. Approximate long-distance recovery sites of Kelp Gulls (*Larus dominicanus*) banded as nestlings near Palmer Station. Not included on the map is the site of the first recovery for no. 0877-34098, because the site location in Argentina remains unknown.

tralia, Tasmania, and New Zealand and its sub-antarctic islands. *L. d. vetula* breeds along coasts of southern Africa.

Current Status in Palmer Archipelago. Kelp Gulls were one of the more abundant and uniformly distributed birds throughout Palmer Archipelago. Although prone to follow ships, they generally remained within sight of land when at sea. Adults and imma-

tures of all ages occurred year-round, but outside of the Palmer Station area their movements are not well documented for the winter months. Fledged birds of the year flocked in the natal areas; some remained in the region while others migrated north to South America, where virtually nothing is known of their duration, movements, food habits, and relations with local populations of the species. The breeding population that occupied Palmer Archipelago was linked closely to the distribution and abundance of Antarctic Limpets (*Nacella concinna*). Most of the gull observations in this study were made by Maxson and Bernstein (1984) and Fraser (1989), who also studied gull-limpet relationships.

Banding. A total of 647 Kelp Gulls were banded within the Palmer Station study area, and an additional 14 were banded on islands close by.

Adults. A total of 121 adults of uncertain age were banded, including 52 at Palmer Station, 49 at Bonaparte Point, 17 at Old Palmer, and 3 at Torgersen Island. Nearly half were banded in 1975-76 (48 birds) by WRF, who also color coded and sexed those individuals studied in detail at Bonaparte Point. SJM and NPB banded an additional 54 adults during 1979-80. Six

adults banded at Bonaparte Point during 1974-75 were recaptured and released there during 1978-79. One adult male banded at Bonaparte Point during 1975-75, and another during 1978-79, were shot there by Hoberg (1983) during 1982-83. To date there have been no recoveries beyond Anvers Island.

Nestlings. A total of 526 nestlings were banded within the Palmer study area, and an additional 3 were banded at Dream Island and 11 at the Joubin Islands. To date there have been six recoveries at foreign sites, all near the Atlantic sector of southern Argentina and Chile (Figure 18 Tables 21, 22). These records indicated that some birds of the year crossed the Drake Passage and spent their first winter in South America. Others ranging in age from two to five years also occurred there at various times.

Surprisingly few nestlings returned to the Palmer natal grounds as breeding adults. Two notable exceptions were ones shot by Hoberg (1983) in search of parasites: 877-34097, banded at Bonaparte Point on 13 January 1975, was collected there on 4 January 1983; 877-34319, banded at Stepping Stones on 19 January 1975, was collected at Bonaparte Point on 5 January 1983. The eight-year-olds were males with brood patches; they were presumably breeding.

Summary Records. Kelp Gulls were widely dispersed throughout Palmer Archipelago. They often were conspicuous where other species were not. Although the gulls bred in many areas, no concerted effort was made to document every nesting at the many places where pairs were observed because of time constraints. Also, the number of breeding pairs often fluctuated dramatically between seasons at any one site. Breeding grounds used fairly frequently within the Palmer study area were Bonaparte Point (high of 81 chicks banded in 1975), Laggard Island (75 chicks in 1984), Stepping Stones (59 chicks in 1975), Limitrophe Island (30 chicks in 1984), and Norsel Point (26 chicks in 1984), but even these areas had few pairs some years. Small numbers usually nested on Elephant Rocks, Litchfield, and DeLaca islands, but elsewhere nesting was highly sporadic, and none was recorded for Torgersen and Christine islands.

The following nestings were recorded beyond the study area in Palmer Archipelago:

Anvers Island. At Bonnier Point (64°27' S, 63°53' W), 1 fledged and 1 nearly fledged chick on 3 February 1979 (DFP, SJM, NPB). At Joubin Islands, 2 fledglings on 5 February 1979 (DFP), 11 large chicks banded on 12 January 1984 (DFP, CCR), and 2 large chicks on 31 December 1984 (DFP, JMP). At Dream Island, 3 large chicks banded on 7 January 1984 (DFP, CCR).

Wiencke Island. At Port Lockroy, nest with two eggs on 28 December 1983; two nests with one and two fresh eggs, respectively, on 17 December 1984; and five fledglings on 26 January 1989 (DFP).

Melchior Islands. Nest with three chicks on 30 December 1983 (CCR), and numerous fledged young on 3 March 1989 (DFP).

Brabant Island. At 64°28' S, 62°34' W, 7 incubating adults on 29 December 1983; at Buls Bay, 2 incubating adults; and at Le Cointe Island, 17 incubating adults on 30 December 1984 (Parmelee and Rimmer 1985).

Liège Island. Numerous fledglings on 31 January 1985 (DFP, MRF).

Winter Records. Holdgate (1963) concluded that the winter population of gulls at Arthur Harbor generally fluctuates according to weather, distribution of open water, and the availability of food around human encampments. Throughout the winter of 1976, Fraser (1989) observed a large and fairly stable population of territorial gulls at Arthur Harbor when ice conditions favored

Table 21. Long-distance recoveries of Kelp Gulls (*Larus dominicanus*) banded as nestlings near Palmer Station

Band no.	Banding date	Banding locality	Recovery date	Recovery locality
0877-34098 (1)	14 Jan. 75	Bonaparte Point	Oct. 80	Argentina, site unknown
0877-34058 (2)	16 Jan. 75	Elephant Seal Rocks	29 Aug. 75	Chile, Tierra del Fuego, 10 km S Cullen 53°00' S, 68°50' W
0877-34316 (3)	19 Jan. 75	Stepping Stones	20 Aug. 76	Argentina, Rio Gallegos
0877-37102 (4)	24 Jan. 77	Bonaparte Point	July 77	Chile, EST Brazo Norte 52°00' S, 70°00' W
1057-10034 (5)	22 Jan. 79	Litchfield Island	31 May 82	Chile, Tierra del Fuego 53°10' S, 70°20' W
1057-10189 (6)	17 Jan. 80	Shortcut Island	27 Aug. 83	Argentina, Rio Grande

Table 22. Long-distance recovery rates of Kelp Gulls (*Larus dominicanus*) banded as nestlings near Palmer Station

Year banded	No. banded	Recovered No.	(%)	Year(s) recovered
1975	183	3	(1.64)	1975, 1976, 1980
1976	57	0	(0)	
1977	19	1	(5.26)	1977
1978	1	0	(0)	
1979	17	1	(5.88)	1982
1980	43	1	(2.33)	1983
1981	6	0	(0)	
1984	214	0	(0)	
Total	540	6	(1.11)	

limpet availability. By contrast, Bernstein (1983) observed a much smaller, unstable population of gulls there during the winter of 1979, when ice conditions forced the limpets to move to deep waters. Fraser concluded that low limpet availability correlated with the presence of pack ice, which, when driven hard against the shore by tides and wind, creates ice-scour conditions highly unfavorable to limpets and other invertebrates. Limpets are favored not only by open water, but also by winter fast ice, which protects the littoral life zone from the movements of pack ice and permits limpet recolonization. Compared to the winter of 1976, that of 1979 was mild and characterized by unstable pack ice conditions, which resulted in low limpet availability and a scarcity of gulls. During the winter of 1983 Heimark and Heimark (1984) witnessed ice conditions and gull numbers at Arthur Harbor similar to those of 1979.

Mild temperatures and unstable ice conditions are common to the seas off Anvers Island (Shabica 1976) and, therefore, a reduced and unstable winter gull population is predictable for most austral winters at Palmer, provided that uncontrolled dumping of refuse does not introduce unnatural food supplies. So sensitive were the gulls to human-produced foods that Bernstein (1983) increased their num-

bers during those periods when he baited nets for trapping and banding purposes. The sudden fluctuations in gull numbers, whether in response to ice shifts or unnatural foods, raise the question as to where the gulls were when they were not at Palmer. Since their departures and returns were often of a sudden nature, it seems probable that the birds simply flew to the nearest open water.

Pietz and Strong (1986) consistently recorded Kelp Gulls between 25 August and 22 September 1985 along transects extending from 61° to 67°30' S, including several runs through Palmer Archipelago. Their report indicates that the species is widespread throughout the region in winter, not necessarily confined to the vicinity of stations. Pietz and Strong (unpublished data) also recorded the presence of immature gulls in larger numbers than heretofore were reported. Flocks of 30 and 21 immatures in company with a few adults, for example, were noted at Arthur Harbor on 30 August 1985. Large numbers, mostly immatures, were also recorded at the Joubin and Gossler islands on 14 September 1985. Immatures also occurred in half of the 40 survey transects in which they observed gulls.

Immatures of all ages occurred at various times throughout winter. Heimark and Heimark (1984) reported birds of the year in first-winter plumage at Arthur Harbor during the winter of 1983: 10 birds on 24 April, 4 on 23 May, 4 on 12 June, 1 on 22 and 25 August, and 7 on 4 September. They recorded second- and third-year birds frequently after 20 September. In 1985 Pietz and Strong (1986) collected the following immature birds: a first-winter male at Fournier Bay, Anvers Island, on 26 August, and a first-winter male and a second-winter female at 64°55.3' S, 64°24.3' W, about 25 kilometers south of Anvers Island on 15 September. A fifth immature in the series was puzzling. The male (collected well south of Anvers Island at 65°21 S, 65°03' W, on 2 September) had a plumage that appeared to be intermediate between that of first- and second-winter birds.

Why some first-year birds migrated to South America and others did not remains a mystery. Fraser (1989) believed that immature gulls had difficulty obtaining limpets and, consequently, migrated to more profitable feeding grounds. Conceivably, the annual production of young influences migration as well. Large numbers of first-year birds flock following a productive season, and they may well be the ones that migrate. The few scattered birds produced during lean years may simply remain on the wintering grounds singly or in small groups associated with adults. Another unsettled question relates to the older immatures residing in South America. Did they migrate there annually, or did they stay until they reached maturity? Did some immatures migrate to South America for the first time after spending their first year or two in Antarctica?

Distribution of Limpets and Gulls.
Although closely related patellid limpets have a circumpolar distribution, *Nacella concinna* is the only species that occupies the islands of the Scotia Arch and the Antarctic Peninsula as far south as Stonington Island, Marguerite Bay (Shabica 1976). According to Fraser (1989), the distribution of Kelp Gulls in Antarctica parallels that of this limpet, which in Palmer Archipelago appears to be the only large and abundant littoral invertebrate among the forty or so species that occur there. Historical records (see Murphy 1936) indicate that formerly throughout their southern and northern ranges Kelp Gulls fed chiefly on intertidal mollusks, but they have since changed their feeding behavior and social regime wherever modern agriculture and industry have had an impact on the envi-

Premigratory flock of immature Kelp Gulls. Although many adult gulls remain in the vicinity of Palmer station throughout the year, most young of the year migrate from the region. Some banded young fly northward to Chile and Argentina. Photographed 9 March 1980 at Hellerman Rocks.

ronment. The intertidal feeding behavior likely will persist in Antarctica only where pristine conditions prevail; its breakdown could be seen readily where open station dumps existed.

Limpet population near Palmer Station.
Stockton (1973) stated that the Antarctic Limpet is more numerous subtidally at Bonaparte Point, but occasionally reaches high densities intertidally, as many as seventy-two individuals having been recorded in an area of 0.025 square meters. In accordance with the findings of other investigators (e.g., Shabica 1976; Picken 1980), Fraser (1989) concluded that the species is characterized by two distinct types showing differences in shell morphology resulting from variation in growth rates rather than genetics. Observed behavioral differences in the two types were attributed to competitive

interactions related to ice and tide conditions. With respect to morphology, one type had a flat, elongate ovate shell (FS) and the other a conical, slightly ovate shell (HS). Fraser stated that the occurrence of FS limpets in the littoral zone at Bonaparte Point was inversely proportional to that of HS limpets. HS limpets were the competitively dominant group that displaced FS limpets during the more equitable, ice-free seasons. Yearly and/or seasonal differences in the littoral proportions of FS limpets were probably influenced by tides, mortality factors, and type and extent of ice cover.

Both types that were eaten by the gulls individually ranged in length from 6 to 60 millimeters. None less than 6 millimeters was found in foods sampled by Fraser. Size

Large numbers of Kelp Gulls occasionally congregate and forage together for limpets. Photographed at Cuverville Island near the Antarctic Peninsula on 18 February 1989.

Shoreline feeding territories. The Palmer study area exemplified the pristine environmental conditions that are necessary for the species' intertidal feeding behavior. Defense of foraging areas by most larid gulls is uncommon, largely temporary, and unrelated to nesting territories. At Palmer, however, the nesting territories of many breeding pairs were adjacent to shoreline feeding territories, a condition not only highly unusual for gulls in general, but also unreported for Kelp Gulls outside of Antarctica.

Fraser (1989) found two distinct types of nesting territories on Bonaparte Point when a near record high of thirty-two pairs bred there in 1975-76. Twenty-two pairs (69 percent) had preferred or "optimal" nesting territories with exclusive access to shoreline foraging areas (feeding territories). Ten pairs (31 percent) used inland ridges without exclusive foraging areas and thus occupied "nonoptimal" territories. The two types of territories showed marked differences. Although each type varied in size, the optimal ones were larger and had lower nesting densities; their feeding territories, however, varied little in size, indicating a uniformly distributed food source. On average, pairs with optimal territories spent more time at the breeding sites, laid their

alone might be the reason the small bivalve mollusk *Kidderia subquadratum* is not an important source of food for the gulls, despite the fact that it is probably the dominant invertebrate at Bonaparte Point. The largest individuals of *subquadratum* measured there by Stockton (1973) were only about 3 millimeters in length.

At Bonaparte Point the gulls were prone to take large limpets of both HS and FS types, but the former were favored when available. Of the two, HS limpets contained the most flesh and consequently yielded the most food with respect to foraging expenditures. According to Branch (1985), the Kelp Gulls at Marion Island in the Indian Ocean not only selected the largest limpets (*Nacella delesserti*), but also those with the palest shells, thus altering the color composition of limpet populations in habitats where predation is most intense.

eggs earlier, laid more large (three-egg) clutches, and quite predictably produced a disproportionately greater number of chicks.

For reasons not fully understood, defense of feeding areas broke down on occasion. Large numbers of gulls congregated and foraged together for limpets in what appeared to be a harmonious communal effort. Fraser (personal communication) thought that such sessions occurred during peaks in limpet availability. The birds were so busy gathering in the bumper crop that no effort was devoted to defense. This behavior was similar to that shown by the gulls at sea when they gorged on swarms of krill. Not only gulls, but seals, skuas, and other seabirds joined the krillfest.

Foraging Behavior. The gulls employed two distinct methods for capturing limpets and, at times, males used a different foraging strategy from that of females. The following observations were made by Fraser (1989), who studied the species' foraging behaviors continuously at Bonaparte Point from 30 October to 7 December 1975, and at intervals from that time through 20 October 1976.

Daily Trends. Littoral exposure at Arthur Harbor occurred twice a day, with optimal foraging at low tide levels. Typically, nesting gulls gathered on the shores of their feeding territories at high tide, where males especially assessed limpet availability during a "search/pause" session. Numbers of gulls increased in the feeding territories as the tides ebbed. Increase in foraging at medium tides was due to greater male participation with less pause time; at low tides the increase resulted from an influx of females. After the chicks hatched, the behavior of the sexes was similar, suggesting that females conserve energy during the egg-laying period not only through courtship feedings, but also by foraging at optimal times.

Foraging and handling techniques. Foraging individuals typically swam slowly and parallel to the shoreline, with heads cocked in a "stiff-necked posture." Experienced adults usually passed up exposed, tightly attached limpets for the more detachable ones that were submerged and active, especially those within a half meter of the water surface.

The gulls captured limpets by two means: (1) "surface seizing" from a floating position that allowed the gulls to reach prey no deeper than 30 centimeters (half body length), and (2) "surface plunging" that was preceded by a very short upward flight followed by a plunge that permitted capture at a depth approaching 70 centimeters (full body length)—nearly twice the depth recorded by Branch (1985) at Marian Island, where Kelp Gulls evidently cannot forage for limpets in water more than 40 centimeters deep. The gulls selectively took limpets of different sizes according to tide levels. The preferred large limpets were captured by surface plunging mostly at high and low tides, and their contents eaten individually on land, where the empty shells were more or less dispersed. Small limpets were captured by surface seizing mostly during medium tides; the gulls swallowed them whole while swimming and later regurgitated their shells in small piles on land. Both large and small shells were deposited within the breeding territories during nesting. Accumulations of shells over time resulted in the formation of large middens, which characterize the species' breeding grounds in Antarctica.

Limpet availability apparently is a function of how fast different-sized individuals move in relation to the speed with which water levels drop. When ebb current pressure began at high tide, large limpets moved quickly into deeper water, leaving behind the smaller and slower limpets. The latter became particularly vulnerable at medium tides, when large limpets were mostly unavailable. Dropping water levels eventually overtook some of the large limpets, and their intake by the gulls increased accordingly. Despite the fact that large limpets yield the most energy for the cost of procurement, and were selected when the gulls had a choice, the gulls took advantage of all limpet sizes and foraging opportunities.

Annual Diet. Kelp Gulls are opportunistic feeders that under pristine conditions in Antarctica prey chiefly on a variety of invertebrates and several species of fish. With respect to the 1975-76 austral season, Fraser (1989) stated that Kelp Gulls fed most frequently on limpets, although their intake varied in response to ice conditions and reproductive events. Food intake during the month of May consisted mostly of the amphipod *Nototropis*, a food usually passed up by the gulls when others were available, according to Fraser, who concluded that limpets and other prey must have been scarce. With the development of fast ice and the littoral recolonization by limpets by June, the diet of the gulls swung back to that prey (85 percent of samples) and continued as such through July and August and into the next breeding season. The limpets were accessible to the gulls by means of open leads that developed between the shorelines of feeding territories and the fast-ice edge. Conversely, the fast ice that promoted limpet availability had

the reverse effect on other prey gleaned from surface water, and probably accounted for the low numbers of these other invertebrates in the foods sampled. Dietary fluctuations can be expected at Arthur Harbor, with its ever-changing ice conditions.

During courtship and egg-laying (September to mid-December), limpets occurred in 84 percent of Fraser's diet samples, no doubt influenced by the rise in limpet availability and courtship feedings involving that particular prey within the territory. During this period, ice conditions had little apparent influence on limpet intake, but they had a decided impact on the lesser food taken, including the fishes *Trematomus bernachii* and *Pleuragramma antarcticum*, as well as the paraphaeid *Pasiphaea longispina*, krill *Euphausia superba*, and amphipod *Nototropis* sp.

At the onset of chick rearing in mid-December there was a noticeable drop in limpet intake. Only 53 percent of Fraser's diet samples taken from January to May contained that prey item. Chicks were fed *Pleuragramma* almost exclusively (98 percent of 396 observed feedings). Adults also fed heavily on this fish; krill and especially paraphaeids were far less important foods to the gulls at this time. Although it was not determined whether fish were more nutri-

tious to chick growth than invertebrates, chick production suffered during those seasons when *Pleuragramma* were not available due to ice cover or other causes.

Despite the importance of fish in the rearing of Kelp Gull chicks, the overall importance of limpets in the species' year-long diet cannot be overstated. Fraser (1989) concluded that clutch size and ultimately the number of fledglings was determined primarily by the ability of males to nourish their mates early in the breeding season. Kelp Gulls nevertheless exploited a seasonally superabundant oceanic prey in fish while feeding chicks, and then switched to a more predictable intertidal prey in limpets at other times of the year. A similar pattern holds for other large larids in polar environments (Ingolfsson 1976; Trapp 1979; Irons et al. 1986). In the Antarctic environment, limpets are not only critical to the courtship and egg-laying periods of the reproductive cycle, but also carry the gulls through periods when other prey may be scarce or unobtainable. Thus limpets make year-long residency possible for the gulls. The lack of a critical number of limpets may be reason enough for gulls to pass up some potential breeding sites.

The long-term study at Palmer Station disclosed the fact that gull numbers fluctuate dramatically at even the traditional breeding sites. Bonaparte Point, for example, had many breeding pairs in the early and mid-1970s. A gradual attrition of pairs from that time until the present led to the widespread belief that human intrusion resulted in the decrease in gull numbers, since that area in particular was frequently visited by station personnel and scientists alike. Fraser (personal communication) believed that the underlying cause was likely a natural local decline in the limpet population on Bonaparte Point; when limpets rebuild their numbers, the gulls presumably will return. His hypothesis merits consideration, because other areas in the Palmer study area have had similar declines, notably the little visited Stepping Stones in Biscoe Bay. Still other areas, such as the lichen-covered cliffs of Hermit Island, with their immense quantities of expended limpet shells, indicate incredible past activity whether due to unusual gull/limpet densities or long periods of normal densities.

Fraser (1989) concluded that early nesting by the Kelp Gulls was influenced by the scarcity of suitable nest sites. Presumably early nesters have the advantage of occupying and retaining the choice sites not only from members of their own kind, but particularly from the migratory South Polar Skuas that arrive and breed late. Early nesting probably has another advantage

with respect to skuas. Predation of Antarctic Tern eggs and chicks increases after the hatching of skua young, when food is most critical for those predators (Parmelee 1977a). Gull chicks from early hatchings probably are less vulnerable to such skua predation than those that hatch late.

Outside of the Palmer study area there were few observations on foods utilized by Kelp Gulls. Thirteen specimens collected by Pietz and Strong (1986) during the winter of 1985 revealed a variety of food items (Table 23). These birds were taken on sea ice or at the edge of sea ice at about 1 to 30 kilometers from land; presumably they had been picking up prey items (invertebrates, fish) and scavenging (birds and mammals) away from their traditional feeding territories. Gulls were not observed to utilize seal carcasses within or near the Palmer study area. However, Pietz and Strong (unpublished data) saw several adults and immatures at seal carcasses on the sea ice well south of Anvers Island on 10 and 12 September 1985. In a wide-open sea southwest of Deception Island in the South Shetlands, DFP noted two adult Kelp Gulls that fed on a floating whale carcass in company with many Southern Giant Petrels, Cape Petrels, and Wilson's Storm-Petrels on 3 January 1978.

Table 23. Food items taken from Kelp Gulls (*Larus dominicanus*) collected by Pietz and Strong (1986) during the winter of 1985

Date	Locality	Sex	Age	Diet
26 Aug. 85	64°32′ S, 63°02′ W	male	adult	shell bits
26 Aug. 85	64°32′ S, 63°02′ W	female	second year	fish (3 otoliths)
26 Aug. 85	64°32′ S, 63°02′ W	female	first year	crustacean parts shell bits
29 Aug. 85	64°59′ S, 63°22′ W	male	adult	shell bits
02 Sept. 85	65°21′ S, 65°03′ W	male	first year	bird (1 quill) flesh (?)
02 Sept. 85	65°21′ S, 65°03′ W	male	adult	crustacean parts
02 Sept. 85	65°21′ S, 65°03′ W	female	adult	unidentified
06 Sept. 85	66°11′ S, 67°47′ W	female	adult	*Euphausia superba* (11 adults, 40 juveniles) fish (vertebrae) seal (hair)
08 Sept. 85	65°35′ S, 64°46′ W	male	adult	squid beaks (16 fragments) crustacean parts bird (12 pinfeathers)
09 Sept. 85	65°08′ S, 65°48′ W	male	adult	fish (3 bodies)
11 Sept. 85	65°04′ S, 65°42′ W	female	adult	squid (1 beak) crustacean parts
15 Sept. 85	64°55′ S, 64°24′ W	female	second year	bird (feathers) fish (vertebrae)
15 Sept. 85	64°55′ S, 64°24′ W	male	first year	bird (feathers)

Kelp Gulls establishing territories on the Bonaparte Point nesting grounds. Photographed 27 October 1975.

Predatory Behavior. The fierce predatory behaviors attributed to many large larids appeared to be mostly lacking in the population of Kelp Gulls inhabiting Palmer Archipelago. In this study there was not a single documented incident of one preying on the eggs or chicks of its own or other species. This unusual gull behavior was obvious from the beginning, when an incubating individual that was closely encircled by several adults suddenly stood up and joined the group and, in so doing, exposed the eggs to the other birds. Not the slightest show of defense was shown by the parent bird, or any apparent interest by the intruders. Although at times the tolerance shown intruding gulls was astounding, the gulls did defend their nest sites forcefully against certain other intruders, especially their archenemy, the South Polar Skua. In defense of its fledglings, one parent gull grasped a skua by the neck in midair and forced it to the ground.

Mortality. Although adult Kelp Gulls appeared to be immune to predatory attacks, their eggs, chicks, and even fledglings were preyed upon by the ever-persistent South Polar Skuas. The gulls were especially sensitive to human intrusion in their nesting areas. Once disturbed, they were reluctant to return to their nests, giving the much bolder skuas an opportunity to plunder. At Palmer a person had to exercise extreme care in dealing with nesting gulls in the presence of skuas, especially when only one parent gull was in attendance. Brown Skuas posed less of a threat, not only because of their small numbers, but also because their territories were apart from those of the gulls. The very abundant South Polar Skuas had their territories adjacent to those of the gulls in many areas. One skua nest was within 5 meters of a gull nest.

Handling young gulls for banding purposes posed another risk. In their attempts to escape, the chicks either hid in rock crevices or swam far out from shore, where they could be retrieved only by boat. Several times Leopard Seals were observed swimming directly past a group of chicks, but although these predators occasionally took penguins in these same waters, they showed not the slightest interest in gull chicks. Another potential predator, the giant petrel, showed disinterest in chicks as well.

Chick mortality due to infanticide, siblicide, or attacks by neighboring gulls was virtually nonexistent, even during periods of food shortage. As a rule, pairs nesting close together were very tolerant of one another's chicks. No fatalities were recorded for chicks trespassing in neighboring territories, unlike other gull species, particularly those in the Northern Hemisphere. Quite the contrary—tolerance increased with the age and growth of the chicks. During the productive years, nurseries with upward of a dozen chicks attended by only one or two adults were commonplace, reminding one of penguin creches or eider nurseries.

Gull mortality was more closely linked to food availability than to inter- or intra-species aggression or predation. So critical are fish in the diets of Kelp Gull chicks that Maxson and Bernstein (1984), as well as Fraser (1989), concluded that limpets obtained from feeding territories adjacent to nest sites are insufficient in themselves to provide enough food for both adults and chicks. Supplementary foods obtained through foraging excursions outside of the territories are indispensable to high breeding success. This was borne out in the fish-poor years at Arthur Harbor, when gull production was also poor, and accounts for the large year-to-year fluctuations in gull production, as is also the case with the fish-dependent South Polar Skuas. A fish-poor year resulted when local fish stocks were low, or when the fish were unavailable due to persistent ice cover or stormy conditions at sea.

Breeding Biology. In addition to his studies on the foraging behavior of Kelp Gulls, Fraser (1989) observed their breeding behavior during the 1974-75 and 1975-76 seasons, when productivity was at a high. By contrast, productivity was at a low point when Maxson and Bernstein (1984) conducted their time-budget studies there in 1978-79 and 1979-80.

Nest sites. Nests were situated on large boulders, rocky prominences, or cliffs within easy access of the sea. Snow-free sites were scarce early in the season and limited the number of nesting pairs. So scarce were sites at Bonaparte Point in 1973 that one nest was built on a snowbank; the nest and its two eggs eventually sank from sight during the subsequent thaw.

Density. Nest patterns were typically colonial, from a few pairs to usually fewer than 50; isolated pairs were uncommon. According to Maxson and Bernstein (1984), nests in high-density areas were within 50 meters of their nearest neighbors, and the pairs had feeding territories less than 100 meters in length.

Site tenacity. Some of the banded pairs at Bonaparte Point refurbished and used their old nests in consecutive years. However, fluctuations in the numbers of breeding pairs in different nesting areas indicated that gulls switch sites, perhaps in response to changes in limpet availability.

Site defense. Intraspecific agonistic behaviors typical of many large larids were not common on the nesting grounds of Kelp Gulls near Palmer Station. These activities accounted for little of the species' time budget throughout its breeding cycle, according to Maxson and Bernstein (1984),

who on occasion observed unusual tolerance by some nesting pairs. In feeding areas, on the other hand, intruding gulls were rarely tolerated; even gulls flying over evoked calls and chases.

Maxson and Bernstein concluded that the intruders were mostly nonbreeding adults or subadults that loafed in high-density areas where agonistic responses largely took place. Males chased intruders away from their nests more often than did females in the high-density areas, but not in the low-density areas. In both areas, males chased intruders on the feeding territory more than did their mates, but females in the high-density areas chased more often than males in the low-density areas.

Other agonistic responses included "face-offs" by males at territorial boundaries, which were infrequent in both high- and low-density areas, and not observed at all in females. "Long calls" often occurred when other gulls flew over the territory. The calls were given most frequently by males in low-density areas, whereas in the high-density areas females long-called more than their mates.

Sexual displays. Fraser (1989) observed that during the winter months pairs frequented and defended their territories, especially the optimal ones. Mewing, choking, head tossing, and other pair-maintenance displays and postures (see Fordham 1963) were common. Courtship feeding of females by males began as early as September in 1975 and 1976 (DRN), and also in 1979 (Bernstein 1983). The time the male spent feeding on territory during the courtship (prelaying) period depended on the availability of certain foods. Because females received most of their food from males during prelaying, their foraging time was independent of available foods and remained fairly constant from one season to the next (Fraser 1989). Maxson and Bernstein (1984) drew similar conclusions from their time-budget studies and stated that females spent less time foraging and more time inactive than their mates. Females frequently begged food from their mates by employing head-tossing displays. Courtship feeding was not observed after completion of the clutch.

Nest-building. According to Fraser (1989), nest-building on Bonaparte Point was essentially continuous from September to March. Maxson and Bernstein (1984) concluded that the peak of active nest-building occurred during the egg-laying period and diminished rapidly once the clutch was complete. Nest construction by both sexes in the beginning was simply the picking away at the centers of old nests.

Later additions were often of material gathered in previous seasons. All attempts at building were halted temporarily following significant spring snows, several of which occurred during 1975. Although females were at the nests for long periods prior to egg-laying, much of their time was spent not in building but in sitting quietly in the empty nest as though incubating (Maxson and Bernstein 1984). From these and other observations by Fordham (1964a) and Fraser (1989), one concludes that males assume the greater responsibility in nest construction.

Some of the nests were hardly more than shallow scrapes; others were large and deeply bowled. They were composed chiefly of chunks of moss, various lichens, particularly the Antarctic Bearded Lichen (*Usnea fasciata*), and the Antarctic Grass (*Deschampsia antarctica*). Beach litter (algae, limpet shells, and so on) also entered into the construction. Most nests were conspicuous and easily seen from afar.

Egg-laying. Egg-laying peaked in late November, and most eggs were laid within a period of one month. Fraser (1989) stated that at Bonaparte Point in 1975, the first and last gull eggs were laid on 18 November and 19 December, respectively. Egg-laying that year probably started even earlier, for six of seventeen nests examined by DFP and DRN on Stepping Stones already had three

eggs by 20 November. Bernstein and Maxson (1984) stated that egg-laying began in mid-November at Bonaparte Point in 197⊆. Heimark and Heimark (1984) recorded th⁻ first eggs there in 1983 on 17 November. The latest egg-laying (back dated from hatching) probably occurred on or after 2⁻ December 1985—a single-egg clutch on Bonaparte Point in an area not known previously to have nesting pairs (DFP).

Clutch size. Clutch size is usually two or three eggs; only two cases of one egg have been recorded. Of 32 pairs that neste⦁ on Bonaparte Point in 1975-76, the averag⦁ number of eggs per pair was approximatel⦁ 2.56 (Fraser 1989).

Incubation period. From laying to hatching of the last egg of the clutch, the period recorded by Fraser (1989) ranged from 26 to 30 days, averaging 28 days for several nests.

Incubation. Maxson and Bernstein (1984) found that although both sexes incubated, females consistently spent more time on the nests than males. Males incubated several hours before midnight, while females typically took the longest shifts during the morning hours, the longest recorded being 10.7 hours. Shorter shifts occurred during the middle portions of the day, when most foraging took place. Because the feeding territories were close by, both parents were able to take advantage of optimal foraging periods by changing over an hour before or after low tide.

Hatching. The observed range in hatching was 5 December–21 January. Peak hatching occurred in late December.

Brooding. By both sexes. According to Maxson and Bernstein (1984), males and females brooded almost equally. Newly hatched chicks were brooded almost continuously, 2-day-olds about 90 percent of the time, and 5-day-olds about 43 percent. Those older than 11 days were no longer brooded. Time spent by both sexes on territory depended on food availability, but generally time off territory increased as the chicks grew and demanded more food.

Prefledging. Maxson and Bernstein (1984) observed that family groups behaved differently during the late chick-rearing stage. Some chicks did not wander far and usually returned to the nest during the evening hours until nearly fledged. Others wandered off when as young as 25 days old and remained well apart from the nest, some being 100 meters away when fledging.

Fledging. The first strong-flying young of the year in the Palmer area were recorded on the following dates: 19 January 1975 (DFP), 31 January 1976 (WRF), 28 January 1977 (DRN), 28 January 1979 (DFP), 10 February 1980 (SJM, NPB), 4 February 1984 (CCR), 15 January 1985 (DFP), and 27 January 1989 (DFP, WRF). Outside of the Palmer area, fledged young were noted at Port Lockroy, Wiencke Island, on 26 January 1989 (DFP).

Fraser (1989) observed that at Bonaparte Point the fledging period ranged from 41 to 47 days, averaging 43 days. One recorded by SJM and NPB was 44 days.

Postfledging period. Parental care continued long after the chicks fledged. Fraser recorded it as late as 1 May, but also noted that 61 fledged chicks had departed Bonaparte Point as early as 5 April in 1976. Maxson and Bernstein (1984) observed chicks being fed nine and fifteen days after fledging; some young continued to beg up to a month after fledging.

Breeding success. The best data relating to breeding success was recorded by Fraser (1989) at Bonaparte Point during the 1975-76 breeding season when productivity peaked. The 32 pairs of gulls that nested there that year laid 82 eggs. Of these, 70 hatched and 12 were lost, mostly to predation by South Polar Skuas, and to a much lesser extent to exposure. Of the 70 chicks, 61 fledged; again, most of the others were lost to predation by South Polar Skuas, but several also to exposure. Fledging success was approximately 1.91 chicks per pair of breeding adults. Lesser losses were attributed to human intrusion and unknown causes. Other seasons were not nearly so productive. Only 2 chicks fledged on Bonaparte Point during the Maxson-Bernstein study, and some years productivity appeared to be zero.

Synopsis of annual cycle. Considering that some Kelp Gulls reside nearly continuously at their nesting/feeding territories, and carry on courtship activities throughout winter, one concludes that the breeding season is continuous year-round in Antarctica, perhaps the last place on earth where one would expect it. On the other hand, parts of the cycle show a remarkable reduction in time when compared with more northerly Kelp Gull populations. During the productive 1975-76 season at Bonaparte Point, Fraser (1989) recorded the first egg on 18 November, and the last fledgling on 28 February—a period of only 102 days. In New Zealand, 147 days elapsed from first egg to last fledging (data from Fordham 1964a, b). This represents a 31 percent contraction in this phase of the breeding cycle for the Antarctic population. One contributing factor is that in Antarctica the gulls are not known to replace lost clutches. Antarctic Terns breeding on the same breeding grounds replaced lost clutches and showed a remarkable protraction of this phase of the cycle.

Antarctic Tern

Antarctic Terns (*Sterna vittata*) were the only species of tern encountered regularly in Palmer Archipelago, where they occurred widely and abundantly throughout the year. Although a few possible sightings of Arctic Terns (*Sterna paradisaea*) were made by members of this study, none was confirmed. One concludes that the species was either a rare or largely overlooked tern whose presence in the archipelago will be documented eventually. The winter appearance of Arctic Terns closely resembles certain immature stages of Antarctic Terns. Generally, Arctic Terns occur at sea in the pack ice; immature Antarctic Terns frequent land areas, often near breeding adults.

Early authorities repeatedly reported the presence of the South American Tern (*Sterna hirudinacea*) at the South Shetlands and other Antarctic islands, but these records were later rejected on the basis of misidentified specimens (Murphy 1936). So far as is known, there is not a valid specimen of *hirudinacea* for this region. A far less likely vagrant, the Bridled Tern (*Sterna anaethetus*), was collected as far south as 56°50' S in the Drake Passage (Peterson and Watson 1971), suggesting that vagrant journeys by *hirudinacea* are quite possible. Another species similar in appearance to *vittata* is the Kerguelen Tern (*Sterna virgata*), but it is known to occur only in the Indian Ocean, where it breeds on Prince Edward, Marion, Crozet, and Kerguelen islands. The highly dissimilar Brown Noddy (*Anous stolidus*) also has a restricted range in the Sub-Antarctic, where it is considered a breeding migrant on the Tristan da Cunha group and Gough Island.

The Antarctic Terns of Palmer Archipelago and adjacent areas are somewhat different from other populations of the species and thought to be a distinct, yet undetermined, race or subspecies (Harrison 1983).

The South Georgia race (*S. v. georgiae*) is noticeably smaller and prone to lay one-egg clutches rather than multiple-egg clutches. *S. v. tristanensis* breeds on Tristan da Cunha and migrates long distances—a distinctly different trait from the year-round population that occupies Palmer Archipelago.

Antarctic Tern
(*Sterna vittata*)

Overall Breeding Distribution. Circumpolar. Breeds in the Maritime Antarctic and all subzones of the Sub-Antarctic. Subspecies *S. v. vittata* breeds on Prince Edward, Marion, Crozet, and Kerguelen islands. *S. v. georgica* breeds on South Georgia. *S. v. tristanensis* breeds on Tristan da Cunha, Gough, St. Paul, and Amsterdam islands. *S. v. bethuni* breeds on sub-antarctic islands of New Zealand. Undetermined subspecies breed on South Orkney and South Shetland islands, and along the Antarctic Peninsula.

Current Status in Palmer Archipelago. Antarctic Terns were abundant year-round residents throughout Palmer Archipelago.

Although they often fed within sight of land, they also occurred far at sea, but were not prone to follow ships. Their breeding grounds frequently shifted from one season to the next, and their breeding cycle was extraordinarily prolonged, considering the polar environment. Most of the tern studies were conducted by Parmelee and Maxson (1974, 1975) and by Parmelee (1977a, 1987, 1988a).

Banding. A total of 202 Antarctic Terns were banded within the Palmer study area: 27 adults of uncertain age and 75 chicks. Ten additional chicks were banded at the Joubin Islands. Nearly all of the adults were banded during 1974-75; most of the chicks were banded during the productive seasons of 1974-75, 1976-77, and 1983-84. To date there has not been a single band recovery. One banded chick later nested at Bonaparte Point on 20 December 1984, but predation of its eggs resulted in site desertion before its band number was determined.

Summer Records. Antarctic Terns were usually encountered at the many places visited in Palmer Archipelago. Most of the larger islands and peninsulas within the study area had small to fairly large colonies from time to time, but for some reason the terns did not use Litchfield, Torgersen,

DeLaca, Christine, and Laggard islands. Their breeding areas frequently shifted no only between seasons but also within them. Relatively few pairs attempted to nest during periods of extensive ice cover.

Nesting terns were noted in many places beyond the study area, but due to time constraints not every encounter was investigated. Not every nest was counted either, because the cryptically colored eggs and chicks were easily stepped on during hasty shore landings. Isolated nestings were not uncommon, but the majority of terns nested in small, rather loose colonies ranging from two to upward of a dozen pairs. Those exceeding forty pairs were exceptional. Colonies recorded outside of the Palmer study area are summarized in Table 24.

Winter Records. Although some Antarctic Terns fly thousands of kilometers from their breeding grounds on Tristan da Cunha in the South Atlantic to the coasts of Africa, where they molt (Cooper 1976), the population that inhabits Palmer Archipelago appeared to be highly sedentary. Since Holdgate's (1963) observations at Arthur Harbor during 1955-57, both adults and immatures have been recorded there repeatedly during the winter months, when their presence or absence depended on the extent of ice cover. With improved ice conditions the birds suddenly returned to Arthur Harbor. It seems probable that they followed a pattern suggested by Watson (1975) in that some Antarctic Terns simply move to the nearest open water rather than to a traditional wintering ground.

With variable ice conditions at Arthur Harbor, their presence in winter was hardly predictable. Heimark and Heimark (1988) noted the species throughout winter in 1983, but not in 1986, when prolonged ice conditions kept the birds away as late as 16 October. Where the birds were then was not determined, but Pietz and Strong (1986) showed that terns were widely dispersed across the region from 62°31' S to 65° S during 22 August–22 September 1985. Their unpublished records included the following locations: Gerlache Strait (seven noted on 25 August); Fournier Bay, Anvers Island (two noted on 26 August); Flandres Bay (three collected on 29 August); Palmer Station to Lemaire Channel near the Antarctic Peninsula (nine noted on 31 August); near Hugo Island at 65°08' S, 65°48' W (five collected on 9 September); Joubin-Gossler-Dream islands (seven noted on 14 September); Deception Island to King George Island in the South Shetlands (twenty-one

Table 24. Approximate locations of Antarctic Tern (*Sterna vittata*) colonies outside of the Palmer study area

Location	Date	Colonies	Observer(s)
Anvers Island			
Joubin Islands	16 Jan. 75	1 large[a]	DFP
Joubin Islands	31 Dec. 84	2 small	DFP, JMP
Paul Island	02 Feb. 79	1 small	DFP, SJM, NPB
Quinton Point	02 Feb. 79	3 small	DFP, SJM, NPB
Bonnier Point	02 Feb. 79	1 small[b]	DFP, SJM, NPB
64°25′ S, 13°53′ W	27 Dec. 83	1 small	DFP, CCR
Gerlache Point	02 Jan. 85	1 small	DFP, JMP
Dream Island	05 Jan. 85	1 small	DFP, JMP
Biscoe Point	28 Jan. 85	1 large	MRF
Wiencke Island			
Port Lockroy	28 Dec. 83	1 small[c]	DFP, CCR
Port Lockroy	29 Dec. 84	1 small	DFP
Port Lockroy	06 Feb. 89	1 small	DFP
Brabant Island			
64°15′ S, 62°32′ W	29 Dec. 83	1 large	DFP, CCR
Metchnikoff Point	29 Dec. 83	1 small	DFP, CCR
Buls Bay	30 Dec. 83	1 small[d]	DFP, CCR
Melchior Islands			
64°19′ S, 62°58′ W	30 Dec. 83	1 small	CCR
Trinity Island			
Skottsberg Point	14 Jan. 84	1 small	DFP
Monument Rocks			
64°00′ S, 60°55′ W	14 Jan. 84	1 small	DFP
Liège Island			
Moureaux Point	31 Jan. 85	1 small	DFP, MRF

Note: Small colonies = 2–12 pairs; large colonies = 13 or more pairs.
[a] Single eggs (5 nests), two eggs (4), egg and chick (1), two chicks (1), scattered chicks (3), one nearly fledged.
[b] Single eggs (2), two eggs (1), single chick (1).
[c] Single eggs (2), two eggs (5).
[d] Single eggs (6), two eggs (5).

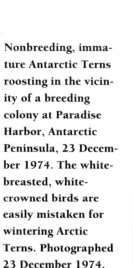

Nonbreeding, immature Antarctic Terns roosting in the vicinity of a breeding colony at Paradise Harbor, Antarctic Peninsula, 23 December 1974. The white-breasted, white-crowned birds are easily mistaken for wintering Arctic Terns. Photographed 23 December 1974.

noted on 20 September). Additional specimens taken over several seasons near Palmer Station proved beyond doubt that both sexes and various age groups were present in winter.

There can be little doubt that Antarctic Terns forage commonly in open water along the ice edge in winter. Among the seven bird species recorded at the ice edge by Fraser et al. (1989), only the Snow Petrels and Antarctic Terns were encountered frequently along the western side of the Antarctic Peninsula during the winter of 1987. These researchers suggested that the

paucity of species along the ice edge in that region typified the winter seabird community elsewhere in Antarctica, based in part on the observations of Griffiths et al. (1982), who considered the wintering avifauna of the African sector to be characterized by relatively few species that occur relatively abundantly.

Fraser et al. (1989) observed that the terns foraged chiefly over open water near the ice edge but, unlike the Snow Petrels, seldom ventured over the ice. And unlike the Snow Petrels, they apparently did not forage at night, thus missing an opportunity to obtain large numbers of krill near the surface during the dark hours.

Immature Terns at Breeding Sites. Conspicuous one- and two-year-old Antarctic Terns with white underparts and black bills frequented the species' breeding grounds and often flocked close to nesting adults. No immature was seen alighting on an empty scrape or abandoned egg, but whenever one of them came too close to a defended territory, it was confronted by the nesting pair and treated like any other intruder. However, the adults occasionally roosted among the immature terns. Their visits to the nesting areas were unpredictable, but nevertheless continued throughout the breeding season. Numbers of immatures usually ranged from a few to upward of a dozen individuals, but as many as thirty-five were recorded one time.

Foraging Behavior and Diet. Antarctic Terns usually foraged fairly close to land, not infrequently within view of their nesting grounds during the breeding season. Simple plunge dives into the sea could be observed almost anytime, but on occasion an individual plunged repeatedly. DRN observed one that plunged eleven times in rapid succession, taking a few seconds between dives for head dunking and feather adjustment. He also noted that the terns took advantage of certain wind and ice conditions that occurred mainly in winter. Whenever strong gales pushed loose brash ice hard against the shore, the terns gathered and, with heads into the wind, feasted on fish and crustaceans brought near the surface during the upheaval caused by churning ice. As noted earlier, Antarctic Terns also occurred commonly along the outer edges of winter pack ice.

At the Snares Islands, New Zealand, Sagar and Sagar (1989) observed that increasing wind speed reduced the terns' capture rate and feeding success. Rough seas significantly reduced the terns' attempt and capture rates. In catching fish, the terns mostly plunge dived in calm seas, but contact dipped for 40-46 percent of attempts in moderate and rough seas. They caught crustaceans mainly by contact dipping, but used partial plunge diving more in moderate and rough seas.

In this study, the only foods observed during feedings or taken directly from the alimentary tracts of specimens were small fish and krill, chiefly *Pleuragramma antarcticum* and *Euphausia superba*. Specimens taken throughout the year (except July) indicated that these were important year-round foods.

Molt. Antarctic Terns residing in Palmer Archipelago underwent a most peculiar molt (Parmelee 1987). Body molt in some adults was under way as early as 19 January; that of the flight feathers began as early as 27 February. Other individuals molted later, likely due to the asynchronous nesting, which resulted in prolonged periods for both breeding and molting. Whether early or late, the molt commenced soon after breeding and transformed the birds into a winter appearance that included dull bills, legs, and feet, as well as paler feathers, particularly of the crown and body underparts. This winter condition was replaced much sooner than previously had been believed, for early molting adults already had the appearance of breeding birds by late May. By late June nearly all adults were in bright breeding condition. No evidence was found that the birds changed from a nonbreeding to a breeding appearance due to a molt of the head and body that reportedly took place during September to December (see Watson 1975). In reality, a rapid molt transformed the birds

Antarctic Tern

Antarctic Terns at nest site on a sea cliff overlooking Buls Bay, Brabant Island, 30 December 1983.

from a breeding appearance to a winter appearance and quickly back to a breeding appearance seemingly in one short, continuous effort. Migrating terns in other parts of the world assume their breeding appearance considerably later, as a result of delaying the molt until after arrival on the wintering grounds (P. G. Ryan, personal communication).

In Palmer Archipelago, one-year-old terns showing traces of the juvenile plumage also molted at different times due to the prolonged fledging period resulting from asynchronous and repeat layings. Specimens showed that the molt in some individuals took place at the time many adults were nesting. One showed body and flight feather molt as early as 25 November, while another showed no molt on 2 December. Still others were in various stages of molt as late as 22 January. Immatures thought to be two-year-olds showed asynchronous molts as well: one specimen in fresh plumage showed virtually no molt as early as 22 April, whereas the remiges of another were short and sheathed as late as ? May. In failing to obtain banded individual of known age, it was not determined when these resident terns first acquired their breeding plumage. Some individuals had the appearance of adults, but on close inspection showed more gray feathers and slightly duller bills and feet than expected for breeding individuals. One such specimen taken from a flock of immatures on 2? December 1974 had undeveloped testes and incoming primary feathers.

Predation and Survival. The Antarctic Tern's long egg-laying period was probably advantageous because the birds were largely inshore feeders with small foraging ranges. According to Croxall and Prince (1980), such an adaptation would reduce intraspecific competition for food at a critical time during chick rearing and adult premolt fattening periods. With respect to South Polar Skua predation, early nesting almost certainly was advantageous, since those terns that fledged early dodged the extra heavy predation that followed increased food demands for those skuas that had chicks. Not all tern colonies were located near skuas, but those that were usually suffered. The many skuas that formed clubs sometimes close to tern colonies were not, however, the ones that were observed harassing the terns. Rather, certain individual skua pairs nesting adjacent to the terns proved to be the principal predators, as described in the section on South Polar Skuas in Chapter 17. Although the fish-dependent skuas could not possibly sustain themselves very long on the limited food provided by a few tern eggs and chicks, for whatever reason certain skuas took a heavy toll. The relatively few Brown Skuas in the study area were not much of a threat to tern survival compared with the nearly ubiquitous South Polar Skuas. Nesting giant petrels and gulls paid little attention to nesting terns, in some cases only 5 meters away. An adult tern accidentally injured near its nest was quickly snatched up by a giant petrel cruising nearby, but only after the cripple had fluttered off and settled in the water some distance from its nest. According to WRF, the surviving parent tern and two large chicks were doing well when last observed eleven days later.

Not all early tern nestings proved advantageous, for following a hard freeze on 22 November 1975 several tern eggs cracked open. Spring snows also forced some of the early layers to abandon their eggs. Colonial nesting would appear to have an advantage over isolated nestings, for the combined forces of many defending terns present a considerable deterrent to predators. However, in this study isolated nestings sometimes succeeded even in those areas where colonies failed. Whether colonial or isolated, those terns nesting well apart from the skuas probably have the advantage, but more observations of this sort are needed. Not observed in the study was loss of nests due to "avalanches of stones," evidently a common cause of mortality in the South Shetland Islands, according to Jablonski (1986).

Winter Fat Reserves. By comparing the weights of a series of adult tern specimens by sex, age, and collection date, it was determined that both males and females weighed significantly more in the winter nonbreeding season than in the summer breeding season (Parmelee 1988a). This increased body weight by fat deposition was thought to be an adaptation to the harsh austral winter rather than simply a recovery due to energetic costs of breeding and molting. The hypothesis was based on the observation that the nonbreeding, relatively inactive immatures that resided near the nesting colonies also weighed significantly more in winter than in summer, indicating that the entire tern population undergoes weight change from one season to the next. The mechanism that controls these changes was not determined, but it appears to be one that turns on fat deposition at the beginning of winter and then turns it off at the start of the breeding season.

Breeding Biology.

Nest sites. In Palmer Archipelago, tern eggs were laid on exposed turf or rock, almost invariably in slight scrapes or natural depressions that often held a few pebbles, bits of vegetation, and especially old limpet shells in which an egg sometimes nestled. As a rule the nests were not far inland, at times near the foot of a glacier, but also commonly on the ledges of precipitous sea cliffs. A few were adjacent to the sea in low-lying areas strewn with polished glacial rubble. In the more hospitable and ice-free areas of the South Shetland Islands, the terns utilized the pebbly beaches above high tide, as well as old beach terraces and even high inclines well back from the sea.

Density. Nesting was typically colonial, from a few nests to usually fewer than 40 pairs, but isolated pairs were not uncom-

Antarctic Tern

thought that the terns favored traditional nesting areas and, like the gulls, bred early to obtain choice sites. Additional observations indicated that the terns did not favor any particular breeding ground or site, and in this they resembled the South American Tern (*Sterna hirdinacea*), which is prone to change nesting areas frequently.

At times Antarctic Terns returned to the same peninsular or island breeding ground, but they usually selected new areas in which to nest. Occasional overlapping of old and new sites occurred, but even then the reuse of previously used scrapes was rare. One reused scrape had an abandoned egg. Two years elapsed between occupancies at another. Although the reuse of scrapes was likely more frequent at cliff sites, where space was limited, attachment to any particular scrape appeared to be accidental rather than intentional. Nesting terns also quickly abandoned a breeding area following the loss of their eggs. Within a short time some of these birds (color marked) renested at another island colony some distance away.

Site defense. Antarctic Terns defended their eggs and chicks both inter- and intraspecifically. Agonistic behavior was most evident in those areas where the nests were close together. The defending tern chal-

mon. Nests in dense colonial situations were usually several meters apart, but in a few cases hardly a meter separated them. Generally, nests were more widely dispersed in extensive snow-free areas, such as commonly occur in the South Shetland Islands. The largest colony (about 100 pairs) noted anywhere in this study was in the South Shetlands at Deception Island on 3 January 1978.

Site tenacity. Antarctic Terns frequently changed breeding grounds between seasons. In the early stages of this study, it was lenged its neighbor with threat postures best described as the "Bill-down Threat Display" (Parmelee and Maxson 1974). The individual called loudly while pressing its body forward with tail up and maintained nearly vertical at times while turning its bill down and holding its closed wings low, somewhat away from the body. Its neighbor challenged similarly, and though the threats lasted only a few seconds, they were often repeated. Threatening birds attacked one another on occasion, but no apparent injuries ensued.

Terns flying overhead elicited a similar response from a defending bird on the ground, except that both bill and tail pointed skyward, giving the body a bow-shaped appearance. Sometimes the defender left the ground and gave chase. Aerial combat above a colony was common, whereas prolonged quiet spells were truly exceptional. Not only did the terns interact among themselves, they also chased everything that moved. Predatory skuas were especially targeted, but also chased were such innocuous individuals as storm-petrels and baby gulls. Humans were constantly dive-bombed, but only rarely did the terns strike a blow.

Sexual display. By midwinter, adult Antarctic Terns already had the appearance of breeding birds and behaved as if paired,

indicating that the pair-bond developed early, or possibly was retained between breeding seasons. Simple pursuit flights thought to relate to courtship, possibly pair-bonding, were noted as early as 11 May in 1975 (DRN) and 10 June in 1976 (WRF). Pursuit flights later in the season included food items: during rapid flight, the pursuer (presumably male) with minnow in beak suddenly raised both wings high above the back and held them there briefly before resuming flapping, usually close behind the pursued. Frequent minnow carrying and presentations prior to the egg-laying appeared to be of vital importance in courtship and probably egg development. This behavior was observed throughout the long breeding season, probably because of asynchronous and repeated nestings.

Nest-scraping. DRN noted that potential nesting areas still blanketed by snow were mildly defended in late September by a few pairs long before any eggs appeared. The birds vocalized and circled above their territories, occasionally alighting on snow or the exposed tops of boulders. Pair numbers and agonistic behaviors increased as the season advanced. DFP observed the selection of a nest site by a pair of terns on Bonaparte Point as early as 3 November; not known was whether these were the terns

(unbanded) that had used this same scrape two years previously. One or both members of this pair visited the scrape sporadically up to the time their egg was laid seventeen days later, on 20 November. Prior to this time the birds had done very little scraping other than pressing flat fresh snow and removing sticks and other small objects placed there intentionally by DFP.

Before settling down to incubate, both sexes at several nests under observation often kicked pebbles out of the scrape with vigorous, deliberate backward thrust of their legs. Incubating individuals frequently picked pebbles, bits of lichens, and soil, and methodically tossed each item to one side or another, also back over the shoulder.

Egg-laying. Antarctic Terns were among the earliest nesters in Palmer Archipelago, as is evidently the case elsewhere in the region (Menegaux 1907; Jablonski 1986). The Second French Antarctic Expedition (Gain 1914) recorded an egg as early as 14 November at Petermann Island (65°10' S, 64°10' W), well south of Anvers Island. The earliest egg date recorded for terns at Palmer was 13 November 1975, but judging by some hatching and fledging dates, it probably commenced even earlier. Peak laying at one colony of thirty-seven nests occurred during 14 and 18 November, but egg-laying for the terns in general continued throughout summer. The latest estimated egg date was 25 February, for DRN observed nestlings at that nest on 27 February 1975, when one of the chicks was probably less than two days old.

Clutch size. One- or two-egg clutches were observed. Not recorded were three-egg clutches, as reported elsewhere for the species, including Petermann Island (Gain 1914). Many one-egg clutches were quickly abandoned for reasons unknown. Considering only those clutches that were being incubated, the ratio of two-egg to one-egg clutches was 27:2 at one colony on Bonaparte Point.

Antarctic Tern chick

Incubation period. From laying to hatching of the last egg of a two-egg clutch, the period twice recorded by DFP was 23 to 24 days. WRF also recorded a 24-day period.

Incubation. Both sexes incubated. During 56 hours of observation at one nest, the female incubated 75 percent of the time. Changeover occurred at irregular times. Prior to changeover, an incubating individual frequently picked and tossed pebbles in the presence of its mate. On occasion it refused to leave even when nudged by the mate. Usually the displaced bird did not leave until after its mate had settled on the eggs. Once incubation was under way the male rarely presented minnows to the female.

Hatching. Two-day intervals occurred between sibling hatchings. The range in hatching dates for a single colony containing thirty-one nests was 6-18 December, but for the entire population, including many colonies, it was much greater (6 December-25 February).

Brooding. Both sexes brooded and fed the chicks, but the male at one closely observed nest did most of the foraging. Brooding adults frequently squirmed and changed positions, with their feathers disarranged and conspicuously puffed out in places.

Feeding of chicks. A brief ceremony often took place upon the arrival of an adult with food for the chick: both parents pointed their bills and tails skyward prior to the moment the food was presented to the brooding bird, which in turn fed the young. Chicks initiated the feeding by first emerging from beneath the brooding parent; failing to do so resulted in aborted feedings. Even the newly hatched chick begged for food by vocalizing and rubbing its mandibles against the bill of its parent. One chick that was only 5 hours old swallowed an entire minnow. If one sibling refused food, it was given to the other, or simply eaten by the parent. Although feedings were irregular, most of the many observed took place within two-hour intervals.

Prefledging. Chicks 60 hours old left the nest voluntarily, but they returned frequently to be brooded, since the parent terns did not brood outside the scrape when the chicks were small. Chicks 72 hours old frequently left the scrape and hid among the rocks whenever the parent flew off. Siblings left the nest finally at nearly the same time, despite their age difference (about 100 and 50 hours, respectively). When not being brooded or fed, the chick hid among the rocks separately or together. Large chicks occasionally scampered over the rocks, but invariably hid in crevices whenever parents protested intrusion. Not determined was the age of the chicks when brooding ceased.

Fledging. Strong-flying fledglings were recorded as early as 29 December 1974 at Bonaparte Point (DFP), and 24 December 1983 at Cormorant Island (DFP, D. Viet). The latest date for large, nearly fledged chicks was 20 March 1975 at Stepping Stones; DRN had observed these chicks there on 27 February when they were probably less than 2 days old. Their fledging period was estimated to be 25 days, as it is somewhat comparable to the incubation period.

Postfledging period. Parental care continued after the chicks fledged, but its duration was not determined. Adults defended fledged chicks near the breeding areas until at least 25 March. DRN observed young terns pursuing flying adults that were carrying fish as late as mid-May.

Synopsis of annual cycle. The year-round resident Antarctic Terns were among the earliest breeders in Palmer Archipelago.

Pair-bonding was already evident by mid-winter, and the birds defended their snow-covered nesting areas long before the eggs appeared. Immatures of all ages flocked at the breeding grounds, which frequently changed locations between seasons. Asynchronous layings and replacement of lost clutches resulted in a prolonged breeding season during which both sexes incubated, brooded, and fed the young. The incubation period was 23-24 days, and the fledging period about 25 days. Viable eggs occurred from at least as early as 12 November to as late as 25 February. Strong-flying young occurred as early as 24 December, and young not quite fledged as late as 20 March. The period from first egg to last fledging was at least 130 days. Ice conditions greatly influenced the species' foraging and reproductive success; South Polar Skuas were its principal predator. Following the breeding season, adult terns underwent a peculiar molt, and their dull winter plumages and other features were short-lived. Both adult and young terns underwent significant seasonal weight changes through fat deposition at the beginning of winter.

Arctic Tern
(*Sterna paradisaea*)

Overall Breeding Distribution. Circumpolar, chiefly in Arctic and Sub-Arctic.

Current Status in Palmer Archipelago. Probably an uncommon but largely overlooked pelagic visitor to Palmer Archipelago. An unsexed specimen (British Museum No. 1949.7.55) was collected at Port Lockroy, Wiencke Island, on 27 February 1945 by personnel of the Falkland Island Dependency. Beyond Palmer Archipelago, two specimens of uncertain sex were collected by G. M. Watson at Harmony Cove, Nelson Island, South Shetland Islands, on 2 March 1964. Southwest of Palmer Archipelago, Arctic Terns were the third most abundant of the 10 species of birds observed from 70°15' S, 88°02' W, to 72°07' S, 135°42' W, in the Bellingshausen and Amundsen seas from 23 January to 15 February 1972 (Erickson et al. 1972).

APPENDIX

Coastal Observations beyond Palmer Study Area

PALMER ARCHIPELAGO

Doumer Island

09 Nov. 74   RRS *John Biscoe* shore landing on northeast coast; also 10 November. DFP.

29 Dec. 84   Coast Guard C-3 survey boat coastal survey of eastern (Peltier Channel) and southern shores. DFP, JMP.

Joubin Islands

16 Jan. 75   RV *Hero* shore landing. DFP, WRF, DRN.

23 Jan. 75   USCGC *Glacier* helicopter landing. DFP, DRN.

05 Feb. 75   RV *Hero* shore landing. DFP, SJM, NPB.

14 Feb. 81   RV *Hero* shore landing. PJP.

12 Jan. 84   Coast Guard C-3 survey boat shore landing. DFP, CCR.

31 Dec. 84   Coast Guard C-3 survey boat shore landing. DFP, JMP.

Wiencke Island

10 Nov. 76   RRS *John Biscoe* shore landing at Damoy Point (64°49' S, 63°32' W). DFP.

28 Dec. 83   RV *Hero* shore landing at Port Lockroy (64°49' S, 63°30' W). DFP.

17 Dec. 84   Coast Guard C-3 survey boat landing at Port Lockroy. DFP, JMP.

29 Dec. 84   Coast Guard C-3 survey boat coastal survey of southwest coast (Peltier Channel) with shore landing at Port Lockroy. DFP, JMP.

05 Jan. 89   MV *Illiria* shore landing at Port Lockroy; also 15 and 26 January, 6 and 19 February, 2 March. DFP, JMP.

10 Jan. 90   MV *Illiria* shore landing at Port Lockroy; also 19 January and 7 February. DFP, JMP.

Dream Island

13 Jan. 77   RV *Hero* shore landing. DFP, DRN, NPB.

06 Feb. 79   RV *Hero* shore landing. DFP, SJM, NPB.

07 Jan. 84   Coast Guard C-3 survey boat shore landing. DFP, CCR.

12 Dec. 84   Coast Guard C-3 survey boat shore landing. DFP, JMP.

05 Jan. 85   Coast Guard C-3 survey boat shore landing. DFP, JMP, MRF.

23 Jan. 85   Coast Guard C-3 survey boat shore landing. DFP, JMP, MRF.

Melchior Islands

25 Dec. 77   RV *Hero* anchorage; also 26 December. DFP, NPB, WL.

01 Feb. 77   RV *Hero* anchorage; also 2 February. DFP, SJM, NPB.

30 Dec. 83   RV *Hero* anchorage; also 31 December. DFP, CCR.

09 Dec. 84   USCGC *Glacier* helicopter aerial survey. DFP.

03 Mar. 89   MV *Illiria* inshore zodiac survey. DFP, JMP.

Anvers Island and small adjacent islands

01 Feb. 79   RV *Hero* coastal survey from Palmer Station along southwest, west, north, and northeast coasts to anchorage at Melchior Islands, with shore landing at 64°14' S, 63°23' W, on north coast. DFP, SJM, NPB.

02 Feb. 79   RV *Hero* coastal survey from Melchior Islands along northeast, north, and west coasts, with shore landings near Cape Gronland (64°16' S, 63°10' W) and Quinton Point (64°20' S, 63°38' W). DFP, SJM, NPB.

03 Feb. 79   RV *Hero* coastal survey along west coast, with shore landings near Bonnier Point (64°27' S, 63°53' W) and Gerlache Point (64°30' S, 63°53' W). DFP, SJM, NPB.

05 Feb. 79   RV *Hero* inshore zodiac survey of southwest coast and Gossler Islands. DFP, SJM, NPB.

27 Dec. 83   RV *Hero* coastal survey of west coast as far north as 64°20' S, 63°41' W, with shore landing at 64°25' S, 63°53' W. DFP, CCR.

08 Dec. 84   USCGC *Glacier* helicopter aerial survey along southwest and west coasts north to Hamburg Bay (64°30' S, 63°57' W). DFP, JMP.

09 Dec. 84   USCGC *Glacier* helicopter aerial survey of Fournier Bay (64°31' S, 63°06' W), Lapeyrere Bay (64°23' S, 63°15' W), and Melchior Islands. DFP.

16 Dec. 84   Coast Guard C-3 survey boat coastal survey of southwest coast to Cape Monaco (64°43' S, 64°18' W). DFP, JMP

17 Dec. 84   Coast Guard C-3 survey boat coastal survey of southeast coast, with shore landing at Biscoe Point (64°49' S, 63°49' W). DFP, JMP.

21 Dec. 84   Coast Guard C-3 survey boat landing at Biscoe Point. DFP, JMP.

02 Jan. 85   Coast Guard C-3 survey boat landing at Gerlache Point. DFP, JMP.

29 Aug. 85   *Polar Duke* WinCruise I visited Fournier Bay. PJP, CSS.

Brabant Island and small adjacent islands

29 Dec. 83   RV *Hero* coastal survey of west coast from 64°30' S, 62°38' W, to 64°03' S, 62°34' W, with shore landings at 64°15' S, 62°32' W, Astrolabe Needle, Guyou Bay, and Metchnikoff Point. DFP, CCR.

30 Dec. 83   RV *Hero* coastal survey of south, east, north, and northwest coasts from 64°31' S, 62°35' W to Metchnikoff Point, with shore landings at Buls Bay, and coastal survey of Grand, Buls, Hunt, Lecointe, Harry, and Davis islands. DFP, CCR.

Trinity Island

14 Jan. 84   RV *Hero* inshore zodiac survey along southwest coast from Skottsberg Point (63°55' S, 60°49' W) to Farewell Rock (63°52' S, 61°01' W). DFP.

16 Jan. 89   MV *Illiria* shore landing at D'Hainaut Island (63°52' S, 60°47' W), Mikkelsen Harbor. DFP, JMP.

Hoseason Island

31 Jan. 85   *Polar Duke* east coast survey from Angot Point (63°48' S, 61°41' W) to Cetacea Rocks (63°43' S, 61°40' W). DFP, JMP, MRF.

Christiana Islands

31 Jan. 85   *Polar Duke* coastal survey of several islands. DFP, JMP, MRF.

Two Hummock Island

30 Jan. 85   *Polar Duke* coastal survey at Cape Kaiser (64°14' S, 62°01' W). DFP, JMP, MRF.

Liège Island

31 Jan. 85   *Polar Duke* east coast survey from Macleod Point (64°06' S, 61°57' W) to Neyt Point (63°58' S, 61°48' W) and inshore zodiac survey along north and northwest coasts from Neyt Point to unnamed bay several nautical miles east of Yoke Island (63°58' S, 61°56' W). DFP, JMP, MRF.

18 Nov. 73   RV *Hero* shore landing on King George Island at Stigant Point (62°02' S, 58°45' W); also 19 November. DFP, SJM.

19 Nov. 73   RV *Hero* shore landing on Nelson Island at Harmony Cove (62°19' S, 59°12' W); also 20 November. DFP, SJM.

13 Dec. 73   RV *Hero* shore landing on Deception Island. DFP, SJM.

13 Dec. 73   RV *Hero* shore landing on Livingston Island at False Bay (62°43' S, 60°22' W). DFP, SJM.

13 Dec. 73   RV *Hero* shore landing on Nelson Island at Harmony Cove; also 14 December. DFP, SJM.

19 Dec. 74   *Lindblad Explorer* shore landing on Elephant Island at Walker Point (61°08' S, 54°42' W). DFP.

21 Dec. 74   *Lindblad Explorer* shore landings on King George Island at Fildes Peninsula (62°12' S, 58°58' W) and Potter Cove (62°14' S, 58°42' W). DFP.

27 Jan. 75   USCGC *Glacier* helicopter landing on Deception Island. DFP.

28 Jan. 75   USCGC *Glacier* helicopter landing on King George Island at Potter Cove. DFP.

01 Dec. 75   RV *Hero* shore landing on Deception Island, also 2 December. DFP, DRN.

30 Dec. 77   RV *Hero* shore landing on Aspland Island (61°30' S, 55°49' W). DFP.

03 Jan. 78   RV *Hero* shore landing on Deception Island. DFP.

14 Jan. 79   RV *Hero* shore landing on King George Island at Admiralty Bay (62°10' S, 58°25' W). DFP, NPB, WL.

09 Nov. 79   RV *Hero* shore landing on King George Island at Admiralty Bay. SJM, PJP, G-AM.

15 Feb. 80   RV *Hero* shore landings on King George Island at Admiralty Bay and Potter Cove. DFP.

15 Feb. 80   RV *Hero* shore landing on Deception Island. DFP.

16 Feb. 80   RV *Hero* shore landing on Snow Island at President Head (62°44' S, 61°12' W). DFP.

13 Mar. 80   RV *Hero* shore landing on Deception Island. DFP, SJM, NPB, PJP, G-AM.

02 Apr. 81   RV *Hero* shore landing on Deception Island. PJP.

15 Jan. 84   RV *Hero* shore landing near King George Island at Ardley Island (62°13' S, 58°56' W); also 16 January. DFP.

01 Feb. 85   *Polar Duke* shore landing on King George Island at Admiralty Bay. DFP, JMP, MRF.

19 Sept. 85   *Polar Duke* shore landing on Deception Island. CSS.

03 Jan. 89   MV *Illiria* shore landing on King George Island at Admiralty Bay; also 25 January and 4 February. DFP, JMP.

07 Jan. 89   MV *Illiria* shore landing on Deception Island; also 16 and 29 January, and 8, 17, and 28 February. DFP, JMP.

07 Jan. 89   MV *Illiria* shore landing on Nelson Island at Harmony Cove; also 4 February. DFP, JMP.

13 Jan. 89   MV *Illiria* shore landing near King George Island at Ardley Island; also 29 January, 8 February. DFP, JMP.

13 Jan. 89   MV *Illiria* shore landing near Livingston Island at Half Moon Island (62°36' S, 59°55' W); also 17 and 28 February. DFP, JMP.

03 Feb. 89   MV *Illiria* inshore zodiac survey on Elephant Island at Cape Lookout. DFP, JMP.

16 Feb. 89   MV *Illiria* shore landing on King George Island at Potter Cove. DFP, JMP.

08 Jan. 90   MV *Illiria* shore landing on King George Island at Admiralty Bay. DFP, JMP.

09 Jan. 90   MV *Illiria* shore landing near Livingston Island at Half Moon Island; also 18 and 30 January, 5 February. DFP, JMP.

09 Jan. 90   MV *Illiria* shore landing on Deception Island; also 18 and 27 January, 5 February. DFP, JMP.

21 Jan. 90   MV *Illiria* shore landing near King George Island at Ardley Island; also 30 January, 8 February. DFP, JMP.

26 Jan. 90   MV *Illiria* inshore zodiac survey on Elephant Island at Cape Lookout; also 4 February. DFP, JMP.

01 Feb. 91   *Polar Circle* shore landing at Deception Island; also 7 February. DFP, JMP, KSW.

01 Feb. 91   *Polar Circle* shore landing at Half Moon Island; also 7 February. DFP, KSW.

ANTARCTIC PENINSULA

21 Nov. 73   RV *Hero* shore landings on Danco Coast at Paradise Bay, including Waterboat Point (64°49' S, 62°52' W). DFP, SJM.

20 Dec. 74   *Lindblad Explorer* shore landing at Hope Bay (63°23' S, 57°00' W). DFP.

22 Dec. 74   *Lindblad Explorer* shore landing at Paradise Bay. DFP.

22 Oct. 75   RV *Hero* shore landing at Paradise Bay. DFP, WRF.

15 Jan. 78   RV *Hero* shore landing at Paradise Bay. DFP.

10 Nov. 79   RV *Hero* shore landing at Paradise Bay. SJM, PJP, G-AM.

01 Apr. 81   RV *Hero* shore landing at Paradise Bay. PJP.

28 Dec. 83   RV *Hero* coastal survey to Argentine Islands (65°15' S, 64°16' W) and Lemaire Channel (65°04' S, 63°57' W). DFP, CCR.

31 Dec. 83   RV *Hero* shore landings on Danco Coast at Orne Harbor (64°37' S, 62°32' W) and Spigot Peak (64°38' S, 62°34' W). DFP, CCR.

14 Jan. 84   RV *Hero* shore landing at Monument Rocks (64°01' S, 60°57' W) off Davis Coast. DFP.

29 Dec. 84   Coast Guard C-3 survey boat coastal survey of Wauwermans Islands (64°55' S, 63°53' W). DFP, JMP.

27 Aug. 85   *Polar Duke* WinCruise I from southern Bransfield Strait as far south as Lavoisier Island (66°12' S, 66°44' W); continuous through 4 September. PJP, CSS.

06 Jan. 89   MV *Illiria* shore landing at Waterboat Point; also 11 January. DFP, JMP.

06 Jan. 89   MV *Illiria* shore landing at Cuverville Island (64°43' S, 62°41' W); also 14 and 26 January, 5 and 18 February, 1 March. DFP, JMP.

28 Jan. 89   MV *Illiria* shore landing at Paradise Bay; also 7 and 19 February, 3 March. DFP, JMP.

05 Feb. 89   MV *Illiria* shore landing at Petermann Island (65°10' S, 64°10' W); also 18 February, 3 March. DFP, JMP.

11 Jan. 90   MV *Illiria* inshore zodiac survey at Paradise Harbor; also 20 January, 7 February. DFP, JMP.

20 Jan. 90   MV *Illiria* shore landing at Waterboat Point; also 29 January. DFP, JMP.

28 Jan. 90   MV *Illiria* shore landing at Cuverville Island. DFP, JMP.

28 Jan. 90   MV *Illiria* shore landing at Petermann Island; also 6 February. DFP, JMP.

06 Feb. 90   MV *Illiria* inshore zodiac survey of Yalour Islands (65°14' S, 64°10' W). DFP, JMP.

28 Jan. 91   *Polar Circle* shore landing at Paulet Island (63°35' S, 55°47' W); also 11 February. DFP, JMP, KSW.

29 Jan. 91   *Polar Circle* inshore zodiac survey at Paradise Bay; also 9 February. DFP, KSW.

30 Jan. 91   *Polar Circle* inshore zodiac survey of Yalour Islands; also 8 February. DFP, KSW.

30 Jan. 91   *Polar Circle* shore landing at Petermann Island. DFP, KSW.

08 Feb. 91   *Polar Circle* shore landing at Argentine Islands (65°15' S, 64°16' W). DFP, JMP, KSW.

11 Feb. 91   *Polar Circle* shore landing at Snow Hill Island (64°28' S, 57°12' W). KSW.

LITERATURE CITED

Ainley, D. G., and S. R. Sanders. 1988. The status of seabirds in the Arthur Harbor/Biscoe Bay area, Antarctica, 1987-88. Report to National Science Foundation, Washington, D.C.

Andersson, K. A. 1905. Wiss. Ergebn. Schwedischen Sudpol. Exped., 1901-1903, Bd. 5 Zool. I, Lief. 2:1-57.

Bagshawe, T. W. 1938. Notes on the habits of the Gentoo and Ringed or Antarctic penguins. Trans. Zool. Soc. Lond. 24:185-306.

Barinaga, M. 1990. Eco-quandary: What killed the skuas? Science 249(4966):243.

Beck, J. R., and D. W. Brown. 1972. The biology of Wilson's Storm Petrel, *Oceanites oceanicus* (Kuhl), at Signy Island, South Orkney Islands. Brit. Antarct. Surv. Sci. Rep., No. 69:1-54.

Beer, C. G. 1970. Individual recognition of voice in the social behavior of birds. Adv. Study Behav. 3:27-74.

Beer, C. G. 1972. Individual recognition of voice and its development in birds. Proc. XV International Ornithological Congress (1970):339-56.

Bennett, A. G. 1922. Breves notas sobre las aves antarticas. Hornero 2:255-58.

Bernstein, N. P. 1982. Activity patterns, energetics, and parental investment of the Antarctic Blue-eyed Shag (*Phalacrocorax bransfieldensis*). Ph.D. thesis, Univ. of Minnesota, Minneapolis-St. Paul.

Bernstein, N. P. 1983. Influence of pack ice on non-breeding Southern Black-backed Gulls (*Larus dominicanus*) in Antarctica. Notornis 30:1-6.

Bernstein, N. P., and S. J. Maxson. 1981. Notes on moult and seasonably variable characters of the Antarctic Blue-eyed Shag *Phalacrocorax atriceps bransfieldensis*. Notornis 28:35-39.

Bernstein, N. P., and S. J. Maxson. 1982a. Behaviour of the Antarctic Blue-eyed Shag (*Phalacrocorax atriceps bransfieldensis*). Notornis 29:197-207.

Bernstein, N. P., and S. J. Maxson. 1982b. Absence of wing-spreading behavior in the Antarctic Blue-eyed Shag (*Phalacrocorax atriceps bransfieldensis*). Auk 99:588-89.

Bernstein, N. P., and S. J. Maxson. 1984. Sexually distinct daily activity patterns of Blue-eyed Shags in Antarctica. Condor 86:151-56.

Bernstein, N. P., and S. J. Maxson. 1985. Reproductive energetics of Blue-eyed Shags in Antarctica. Wilson Bull. 97:450-62.

Bernstein, N. P., and P. C. Tirrell. 1981. New southerly record for the Macaroni Penguin (*Eudyptes chrysolophus*) on the Antarctic Peninsula. Auk 98:398-99.

Branch, G. M. 1985. The impact of predation by Kelp Gulls *Larus dominicanus* on the Sub-Antarctic Limpet *Nacella delesserti*. Polar Biol. 4:171-77.

Bretagnolle, V. 1989a. Temporal progression of the giant petrel courtship. Ethology 80:245-54.

Bretagnolle, V. 1989b. Calls of Wilson's Storm Petrel: Functions, individual and sexual recognitions, and geographic variation. Behaviour 111:98-112.

Burger, A. E. 1980. Behavioural ecology of Lesser Sheathbills *Chionis minor* at Marion Island. Ph.D. thesis, Univ. of Cape Town, South Africa.

Burton, R. W. 1968. Breeding biology of the Brown Skua, *Catharacta skua lonnbergi* (Mathews), at Signy Island, South Orkney Islands. Brit. Antarct. Surv. Bull. 15:9-28.

Carrick, R., and S. E. Ingham. 1970. Ecology and population dynamics of antarctic sea birds. In M. W. Holdgate (ed.), *Antarctic ecology*. London: Academic Press, pp. 505-25.

Carroll, A. M. 1954. Unpublished bird report. Brit. Antarct. Surv. Sci. Rep., No. Q58/1954/A.

Chappell, M. A., and S. L. Souza. 1987. Physiological ecology of Adélie Penguins during the reproductive season. Antarct. J. U.S. 22(5):228-29.

Conroy, J. W. H. 1972. Ecological aspects of the biology of the Giant Petrel *Macronectes giganteus* (Gmelin) in the maritime Antarctic. Brit. Antarct. Surv. Sci. Rep., No. 75:1-74.

Conroy, J. W. H., and E. L. Twelves. 1972. Diving depths of the Gentoo Penguin (*Pygoscelis papua*) and Blue-eyed Shag (*Phalacrocorax atriceps*) from the South Orkney Islands. Brit. Antarct. Surv. Bull. 30:106-8.

Cooper, J. S. 1976. Seasonal and spatial distribution of the Antarctic Tern in South Africa. S. Afr. T. Antarkt., Nov. Deel 6:30-32.

Cramp, S. and K. E. L. Simmons (eds.). 1983. *Handbook of the birds of Europe, the Middle East and North Africa.* Vol. III. *Waders to gulls.* Oxford: Oxford Univ. Press.

Croxall, J. P., and E. D. Kirkwood. 1979. *The distribution of penguins on the Antarctic Peninsula and islands of the Scotia Sea.* Cambridge: Brit. Antarc. Surv.

Croxall, J. P., and P. A. Prince. 1980. Food, feeding ecology and ecological segregation of seabirds at South Georgia. Biol. J. Linnean Soc. 14:103-31.

Daniels, R. A., and J. H. Lipps. 1982. Distribution and ecology of fishes of the Antarctic Peninsula. J. Biogeo. 9(1):1-9.

DeLaca, T. E., and J. H. Lipps. 1976. Shallow-water marine associations, Antarctic Peninsula. Antarct. J. U.S. 11(1):12-20.

Devillers, P. 1978. Distribution and relationships of South American skuas. Le Gerfaut 68:374-417.

Devillers, P., and J. A. Terschuren. 1978. Relationships between the blue-eyed shags of South America. Le Gerfaut 68:53-86.

Eastman, J. T. 1985. *Pleuragramma antarcticum* (Pices, Nototheniidae) as food for other fishes in McMurdo Sound, Antarctica. Polar Biol. 4:155-60.

Eppley, Z. A., and M. A. Rubega. 1989. Indirect effects of an oil spill. Nature 340:513.

Erickson, A. W., J. R. Gilbert, G. A. Petrides, R. J. Oehlenschlager, A. A. Sinha, and J. Otis. 1972. Populations of seals, whales, and birds in the Bellingshausen and Amundsen seas. Antarct. J. U.S. 7(4):70-72.

Falla, R. A. 1964. Distribution patterns of birds in the Antarctic and high latitude Subantarctic. In A. Carrick et al. (eds.), *Biologie Antarctique.* Paris: Hermann, pp. 367-78.

Fogg, G. E., and D. Smith. 1990. *The exploration of Antarctica.* London: Cassel.

Fordham, R. A. 1963. Individual and social behaviour of the Southern Black-backed Gull. Notornis 10:206-22.

Fordham, R. A. 1964a. Breeding biology of the Southern Black-backed Gull I: pre-egg and egg stage. Notornis 11:3-34.

Fordham, R. A. 1964b. Breeding biology of the Southern Black-backed Gull II: incubation and the chick stage. Notornis 11:110-26.

Fraser, W. R. 1989. Aspects of the ecology of Kelp Gulls (*Larus dominicanus*) on Anvers Island, Antarctic Peninsula. Ph.D. thesis, Univ. of Minnesota, Minneapolis-St. Paul.

Fraser, W. R., R. L. Pitman, and D. G. Ainley. 1989. Seabird and fur seal responses to vertically migrating winter krill swarms in Antarctica. Polar Biol. 10:37-41.

Gain, L. 1914. Oiseaux antarctiques. Deuxieme Exp. Antarct. Francaise, 1908-10. Paris.

Glass, B. M. 1978. Winter observations of birds at Palmer Station in 1977. Antarct. J. U.S. 13 (4):145.

Griffiths, A. M., W. R. Siegfried, and R. W. Abrams. 1982. Ecological structure of a pelagic seabird community in the Southern Ocean. Polar Biol. 1:39-46.

Guard, C. L., and D. E. Murrish. 1979. Blood flow in the giant petrel. Antarct. J. U.S. 9 (1):101-3.

Halpryn, B. M., P. C. Tirrell, and D. E. Murrish. 1982. Circadium rhythms in the body temperature of Adélie Penguins. Antarct. J. U.S. 17(5):182-83.

Harrison, P. 1983. *Seabirds.* Boston: Houghton Mifflin.

Heimark, G. M., and R. J. Heimark. 1984. Birds and marine mammals in the Palmer Station area. Antarct. J. U.S. 19(4):3-8.

Heimark, G. M., and R. J. Heimark. 1988. Obsevations of birds and marine mammals at Palmer Station November 1985 to November 1986. Antarct. J. U.S. 23(4): 14-17.

Hemmings, A.D. 1984. Aspects of the breeding biology of McCormick's Skua *Catharacta maccormicki* at Signy Island, South Orkney Islands. Brit. Antarct. Surv. Bull. 65:65-79.

Herwig, R. P., J. Stemmler, K. A. Nagy, G. Garrett, B. S. Obst, C. H. Stinson, J. S. Maki and J. T. Staley. In press. Chitin degradation and energy assimilation efficiency in the Adélie Penguin *Pygoscelis adeliae.* Polar Biol.

Hoberg, E. P. 1983. Preliminary comments on parasitological collections from seabirds at Palmer Station, Antarctica. Antarct. J. U.S. 18(5):206-8.

Holdgate, M. W. 1963. Observations of birds and seals at Anvers Island, Palmer Archipelago, in 1955-57. Brit. Antarct. Surv. Bull. 2:45-51.

Holm-Hansen, O., B. G. Mitchell, C. D. Hewes, and D. M. Karl. 1989. Phytoplankton blooms in the vicinity of Palmer Station, Antarctica. Polar Biol. 10:49-57.

Hopkins, T. L. 1985. The zooplankton community of Crocker Passage, Antarctic Peninsula. Polar Biol. 4:161-70.

Hunter, S. 1983. The food and feeding ecology of the Giant Petrels *Macronectes halli* and *M. giganteus* at South Georgia. J. Zool., London 200:521-38.

Hunter, S. 1984a. Movements of Giant Petrels *Macronectes* sp. ringed at South Georgia. Ringing and Migration 5:105-12.

Hunter, S. 1984b. Breeding biology and population dynamics of giant petrels *Macronectes* at South Georgia (Aves: Procellariiformes). J. Zool., London 203:441-60.

Ingolfsson, A. 1976. The feeding habits of Great Black-backed Gulls, *Larus marinus*, and Glaucous Gulls, *L. hyperboreus*, in Iceland. Acta Naturalia Islandica 24:2-19.

Irons, D. B., R. G. Anthony, and J. A. Estes. 1986. Foraging strategies of Glaucous-winged Gulls in a rocky intertidal community. Ecol. 67:1460-74.

Jablonski, B. 1984. Distribution and numbers of penguins in the region of King George Island (South Shetland Islands) in the breeding season 1980/1981. Polish Polar Res. 5:17-30.

Jablonski, B. 1986. Distribution, abundance and biomass of a summer community of birds in the region of the Admiralty Bay (King George Island, South Shetland Islands, Antarctic) in 1978/79. Polish Polar Res. 7:217-60.

Johnstone, G. W. 1977. Comparative feeding ecology of the giant petrels *Macronectes giganteus* (Gmelin) and *M. halli* (Matthews). In G. A. Llano (ed.), *Adaptations within antarctic ecosystems*. (Proc. Third SCAR Symp. Antarct. Biol. by Smithsonian Institution, Washington, D.C.), pp. 647-68.

Johnstone, G. W. 1979. Agonistic behaviour of the giant petrels *Macronectes giganteus* and *M. halli* feeding at seal carcasses. Emu 79 (3):129-32.

Jones, N. V. 1963. The Sheathbill *Chionis alba* (Gmelin), at Signy Island, South Orkney Islands. Brit. Antarct. Surv. Bull. 2:53-71.

Jouventin, P., and M. Guillotin. 1979. Socio-ecologie du skua antarctique a Pointe Geologie. Terra et Vie, Rev. Ecol. 33:109-27.

Komarkova, V. 1983. Plant communities of the Antarctic Peninsula near Palmer Station. Antarct. J. U.S. 18(5):216-18.

Komarkova, V. 1984. Studies of plant communities of the Antarctic Peninsula near Palmer Station. Antarct. J. U.S. 19(5):180-82.

Lee, R. E., and J. G. Baust. 1982a. Physiological adaptations of antarctic terrestrial arthropods. Antarct. J. U.S. 17(5):193-95.

Lee, R. E., and J. G. Baust. 1982b. Respiratory metabolism of the Antarctic Tick, *Ixodes uriae*. Comp. Biochem. Physiol. 72A:167-71.

Lee, R. E., and J. G. Baust. 1987. Cold-hardiness in the Antarctic Tick, *Ixodes uriae*. Physiol. Zool. 60:499-506.

Matthew, K. 1982. Rockhopper Penguin (*Eudyptes chrysocome*) record at Palmer Station, Antarctica. Auk 99:384.

Maxson, S. J., and N. P. Bernstein. 1980. Ecological studies of Southern Black-backed Gulls, Blue-eyed Shags, and Adélie Penguins at Palmer Station. Antarct. J. U.S. 15(5):157.

Maxson, S. J., and N. P. Bernstein. 1982. Klepto-parasitism by South Polar Skuas on Blue-eyed Shags in Antarctica. Wilson Bull. 94(3):269-81.

Maxson, S. J., and N. P. Bernstein. 1984. Breeding season time budgets of the Southern Black-backed Gull in Antarctica. Condor 86:401-9.

Menegaux, A. 1907. Expedition Antarctique Francaise. 1903-1905. Sci. Nat. Doc. Sci., Oiseaux:1-75.

Mougin, J. L. 1968. Etude ecologique dequatre especes de petrels antarctiques. Oiseau 38 (No. special):1-52.

Mougin, J. L. 1975. Ecologie comparee des procellariidae Antarctiques et Subantarctiques. Comite National Francais des Recherches Antartiques 36:1-195.

Moynihan, M. 1955. Some aspects of reproductive behavior in the Black-headed Gull (*Larus ridibundus* L.) and related species. Behav., Suppl. 4.

Moynihan, M. 1962. Hostile and sexual behavioral patterns of South American and Pacific Laridae. Behav., Suppl. 8.

Muller-Schwarze, C., and D. Muller-Schwarze. 1975. A survey of twenty-four rookeries of pygoscelid penguins in the Antarctic Peninsula region. In B. Stonehouse (ed.), *The biology of penguins*. London: Macmillan, pp. 309-20.

Muller-Schwarze, D. 1984. Possible human impact on penguin populations in the Antarctic Peninsula area. Antarct. J. U.S. 19 (5):158-59.

Murphy, R. C. 1936. *Oceanic birds of South America.* Vol. 2. New York: American Mus. Nat. Hist.

Murrish, D. E. 1982. Acid-base balance in three species of antarctic penguins exposed to thermal stress. Physiol. Zool. 55:137-43.

Murrish, D. E., and C. L. Guard. 1977. Cardiovascular adaptations of the Giant Petrel, *Macronectes giganteus*, to the antarctic environment. In G. A. Llano (ed.), *Adaptations within antarctic ecosystems.* (Proc. Third SCAR Symp. Antarct. Biol. by Smithsonian Institution, Washington, D.C.), pp. 511-30.

Murrish, D. E., and P. C. Tirrell. 1981. Respiratory heat and water exchange in three species of antarctic birds. Antarct. J. U.S. 16(5):148-50.

Nagy, K. A., B. S. Obst, and R. D. Wilson. 1984. Energetics and feeding ecology of free-living penguins and petrels. Antarct. J. U.S. 19 (5):163-64.

Neilson, D. R. 1983. Ecological and behavioral aspects of the sympatric breeding of the South Polar Skua (*Catharacta maccormicki*) and the Brown Skua (*Catharacta lonnbergi*) near the Antarctic Peninsula. M.S. thesis, Univ. of Minnesota, Minneapolis-St. Paul.

Obst, B. S. 1985. Densities of antarctic seabirds at sea and the presence of krill *Euphausia superba*. Auk 102:540-49.

Obst, B. S. 1986. Wax digestion in Wilson's Storm-Petrel. Wilson Bull. 98(2):189-95.

Obst, B. S. In press. Stomach oil and the energetics of Wilson's Storm-Petrel nestlings. Auk.

Obst, B. S., K. A. Nagy, and R. E. Ricklefs. 1987. Energy utilization by Wilson's Storm-Petrel. Physiol. Zool. 60:200-210.

Parker, B. C., G. L. Samsel, and G. W. Prescott. 1972. Freshwater algae of the Antarctic Peninsula: Preliminary observations in the Palmer Station area, Anvers Island, Antarctica. Ant. Res. Series (American Geophysical Union) 18:69-81.

Parmelee, D. F. 1977a. Adaptations of Arctic and Antarctic terns within antarctic ecosystems. In G. A. Llano (ed.), *Adaptations within antarctic ecosystems.* (Proc. Third SCAR Symp. Antarct. Biol. by Smithsonian Institution, Washington, D.C.), pp. 687-702.

Parmelee, D. F. 1977b. Review: "Birds of the Antarctic and Sub-Antarctic" by G. E. Watson. Wilson Bull. 89(4):646-48.

Parmelee, D. F. 1980. *Bird island in antarctic waters.* Minneapolis: Univ. of Minnesota Press.

Parmelee, D. F. 1985. Polar adaptations in the South Polar Skua (*Catharacta maccormicki*) and the Brown Skua (*Catharacta lonnbergi*) of Anvers Island, Antarctica. In V. D. Ilyichev and V. M. Gavrilov (eds.), *ACTA 18 Congressus Internationalis Ornithologici.* Vol. I. Moscow: Nauka, pp. 520-29.

Parmelee, D. F. 1987. Unexpected plumage in Antarctic Terns *Sterna vittata* during the austral winter. Cormorant 15:41-47.

Parmelee, D. F. 1988a. Unexpected weight gain in resident Antarctic Terns *Sterna vittata* during the austral winter. Ibis 130:438-43.

Parmelee, D. F. 1988b. The hybrid skua: a Southern Ocean enigma. Wilson Bull. 100 (3):345-56.

Parmelee, D. F., N. Bernstein, and D. R. Neilson. 1978. Impact of unfavorable ice conditions on bird productivity at Palmer Station during the 1977-78 field season. Antarct. J. U.S. 13 (4):146-47.

Parmelee, D. F., and W. R. Fraser. 1989. Multiple sightings of Black-necked Swans in Antarctica. American Birds 43(4):1231-32.

Parmelee, D. F., W. R. Fraser, B. Glass, and D. R. Neilson. 1977a. Ecological and behavioral adaptations to antarctic environments. Antarct. J. U.S. 12(4):17.

Parmelee, D. F., W. R. Fraser, and D. R. Neilson. 1977b. Birds of the Palmer Station area. Antarct. J. U.S. 12(1, 2):14-21.

Parmelee, D. F., W. R. Fraser, and D. R. Neilson. 1975. Ornithological investigations at Palmer Station. Antarct. J. U.S. 10(4):124-25.

Parmelee, D. F., and S. D. MacDonald. 1973. Birds of the antarctic ice pack. Antarct. J. U.S. 8(4):150.

Parmelee, D. F., and S. D. MacDonald. 1975. Recent observations on the birds of Isla Contramaestre and Isla Magdalena, Straits of Magellan. Condor 77:218-20.

Parmelee, D. F., and S. J. Maxson. 1974. Ornithological investigations of the Antarctic Tern. Antarct. J. U.S. 9(4):103.

Parmelee, D. F., and S. J. Maxson. 1975. The Antarctic Terns of Anvers Island. Living Bird 13:233-50.

Parmelee, D. F., S. J. Maxson, and N. P. Bernstein. 1979. Fowl cholera outbreak among Brown Skuas at Palmer Station. Antarct. J. U.S. 14(5):168-69.

Parmelee, D. F., and J. M. Parmelee. 1987a. Revised penguin numbers and distribution for Anvers Island, Antarctica. Brit. Antarct. Surv. Bull. 76:65-73.

Parmelee, D. F., and J. M. Parmelee. 1987b. Movements of Southern Giant Petrels ringed near U.S. Palmer Station, Antarctica. Ringing and Migration 8:115-18.

Parmelee, D. F., J. M. Parmelee, and M. Fuller. 1985. Ornithological investigations at Palmer Station: The first long-distance tracking of seabirds by satellites. Antarct. J. U.S. 19 (5):162-63.

Parmelee, D. F., and P. J. Pietz. 1987. Philopatry, mate and nest-site fidelity in the Brown Skuas of Anvers Island, Antarctica. Condor 89:916-19.

Parmelee, D. F., and C. Rimmer. 1984. Status of known-age birds banded as chicks near Palmer Station in the 1970s. Antarct. J. U.S. 19(5):164-65.

Parmelee, D. F., and C. C. Rimmer. 1985. Ornithological observations at Brabant Island, Antarctica. Brit. Antarct. Surv. Bull. 67:7-12.

Penhale, P. A. 1989. Research team focuses on environmental impact of oil spill. Antarct. J. U.S. 24(2):9-12.

Peter, H.-U., M. Kaiser, and A. Gebauer. 1988a. Investigations on birds and seals on King George Island (South Shetland Islands, Antarctic). Geodatische and Geophysikalische Veroffentlichungen Reihe I Heft 14:1-127.

Peter, H.-U., M. Kaiser, A. Gebauer, and D. Zippel. 1988b. Zur Dynamik des Winterbestande des Weisgesicht-Scheidenschnabels (*Chionis alba*) auf King George Island (South Shetland Islands). Beitr. Vogelk. 34:205-20.

Peter, H.-U., M. Kaiser, and A. Gebauer. 1990. Ecological and morphological investigations on South Polar Skuas (*Catharacta maccormicki*) and Brown Skuas (*Catharacta skua lonnbergi*) on Fildes Peninsula, King George Island, South Shetland Islands. Zool. Jb. Syst. 117:201-18.

Peterson, R. T., and G. E. Watson. 1971. Franklin's Gull and Bridled Tern in southern Chile. Auk 88:670-71.

Picken, G. B. 1980. The distribution, growth, and reproduction of the Antarctic Limpet *Nacella* (*Patinigera*) *concinna* (Strebel, 1908). J. Exper. Marine Biol. and Ecol. 42:71-85.

Pietz, P. J. 1982. Comparative ecology of South Polar and Brown skuas at Palmer Station. Antarct. J. U.S. 17(5):176.

Pietz, P. J. 1984. Aspects of the behavioral ecology of sympatric South Polar and Brown skuas near Palmer Station, Antarctica. Ph.D. thesis, Univ. of Minnesota, Minneapolis-St. Paul.

Pietz, P. J. 1985. Long call displays of sympatric South Polar and Brown skuas. Condor 87:316-26.

Pietz, P. J. 1986. Daily activity patterns of South Polar and Brown skuas near Palmer Station, Antarctica. Auk 103:726-36.

Pietz, P. J. 1987. Feeding and nesting ecology of sympatric South Polar and Brown skuas. Auk 104:617-27.

Pietz, P. J., and G.-A. D. Maxson. 1980. Comparative ecology of South Polar and Brown skuas at Palmer Station. Antarct. J. U.S. 15(5):158.

Pietz, P. J., and C. S. Strong. 1986. Ornithological observations west of the Antarctic Peninsula, August-September 1985. Antarct. J. U.S. 21(5):203-4.

Poncet, S., and J. Poncet. 1987. Censuses of penguin populations of the Antarctic Peninsula, 1983-87. Brit. Antarct. Surv. Bull. 77:109-129.

Price, D. M. 1959. Unpublished bird report. Brit. Antarct. Surv. Sci. Rep., No. Q1/1959/A.

Quetin, L. B., and R. M. Ross. 1988. The biological basis for area-wide protection at Palmer Station, Antarctica. Unpublished report of a workshop sponsored by the Marine Mammal Commission (Contract No. T6223924-6), 59 pp.

Rasmussen, P. C. 1988a. Moults of rectrices and body plumage of Blue-eyed and King shags (*Phalacrocorax atriceps* and *P. albiventer*) and phenology of moults. Notornis 35:129-42.

Rasmussen, P. C. 1988b. Stepwise molt of remiges in Blue-eyed and King shags. Condor 90:220-27.

Rasmussen, P. C., and P. S. Humphrey. 1988. Wing-spreading in chilean Blue-eyed Shags (*Phalacrocorax atriceps*). Wilson Bull. 100:140-44.

Ricklefs, R. E. 1982. Development of homeothermy in antarctic seabirds. Antarct. J. U.S. 17(5):177-78.

Ricklefs, R. E., and K. K. Matthew. 1983. Rates of oxygen consumption in four species of seabirds at Palmer Station, Antarctic Peninsula. Comp. Biochem. Physiol. 74A:885-88.

Risebrough, R. W., G. E. Watson, and J. P. Angle. 1976. A Red Phalarope (*Phalaropus fulicarius*) in breeding plumage on Anvers Island. Antarct. J. U.S. 11(4):226.

Roberts, B. 1940. The life cycle of Wilson's Petrel *Oceanites oceanicus* (Kuhl). British Graham Land Expedition, 1934-1937, Sci. Rep. 1, 441-94.

Roots, D. M. 1988. The status of birds at Signy Island, South Orkney Islands. Brit. Antarct. Surv. Bull. 80:87-119.

Sagar, P. M., and J. L. Sagar. 1989. The effects of wind and sea on the feeding of Antarctic Terns at the Snares Islands, New Zealand. Notornis 36:171-82.

Salomonsen, F. 1976. The South Polar Skua *Stercorarius maccormicki* Saunders in Greenland. Danks Orn. Foren. Tidsskr. 70:81-89.

Schlatter, R. P., and C. A. Moreno. 1976. Habitos alimentarios del Cormoran Antartico, *Phalacrocorax atriceps bransfieldensis* (Murphy) en Isla Green, Antarctica. Ser. Cient. Inst. Antarct. Chileno 4:69-88.

Schlichting, H. E., Jr., B. J. Speziale, and R. M. Zink. 1978. Dispersal of algae and Protozoa by antarctic flying birds. Antarct. J. U.S. 13 (4):147-49.

Shabica, S. V. 1976. The natural history of the Antarctic Limpet *Patinigera polaris* (Hombron and Jaquinot). Ph.D. thesis, Univ. of Oregon, Corvallis.

Shaw, P. 1981. The breeding biology of the shag *Phalacrocorax atriceps*, 1980/81. Brit. Antarct. Surv. Annual Rep.

Shaw, P. 1985a. Brood reduction in the Blue-eyed Shag *Phalacrocorax atriceps*. Ibis 127:476-94.

Shaw, P. 1985b. Age-differences within breeding pairs of Blue-eyed Shags *Phalacrocorax atriceps*. Ibis 127:537-43.

Shaw, P. 1986. Relationship between dominance behaviour, bill size and age group in Greater Sheathbills *Chionis alba*. Ibis 128:48-56.

Short, L. 1969. Taxonomic aspects of avian hybridization. Auk 86:84-105.

Sickles, S. A., and D. E. Murrish. 1983. The effect of lighting conditions on circadian rhythmicity in the body temperatures of Adélie Penguins. Antarct. J. U.S. 18(5):210-11.

Siegel-Causey, D., and C. Lefevre. 1989. Holocene records of the Antarctic Shag (*Phalacrocorax* [*Notocarbo*] *bransfieldensis*) in Fuegian waters. Condor 91:408-15.

Sladen, W. J. L. 1954. Pomarine Skua in the Antarctic. Ibis 96:315-16.

Smith, R. I. L. 1982. Plant succession and reexposed moss banks on a deglaciated headland in Arthur Harbor, Anvers Island. Brit. Ant. Surv. Bull. 51:193-99.

Smith, R. I. L., and R. W. M. Corner. 1973. Vegetation of the Arthur Harbor—Argentine Island region of the Antarctic Peninsula. Brit. Ant. Surv. Bull. 33, 34:89-122.

Snow, B. K. 1960. The breeding biology of the Shag *Phalacrocorax aristotelis* on the island of Lundy, Bristol Channel. Ibis 102:554-75.

Snow, B. K. 1963. The behavior of the Shag. Brit. Birds 56:77-103, 164-86.

Stemmler, J., R. P. Herwig, J. T. Staley, and K. Nagy. 1984. Chitin degradation in Adélie Penguins. Antarct. J. U.S. 19(5):161-62.

Stockton, W. L. 1973. An intertidal assemblage at Palmer Station. Antarct. J. U.S. 8(5):305-7.

Stout, W. E., and S. V. Shabica. 1970. Marine ecological studies at Palmer Station and vicinity. Antarct. J. U.S. 5(4):134-35.

Strikwerda, T. E., M. R. Fuller, W. S. Seegar, P. W. Howey, and H. D. Black. 1986. Birdborne satellite transmitter and location program. Johns Hopkins APL Tech. Digest 7 (2):203-8.

Tinbergen, J. 1957. Unpublished bird report. Brit. Antarct. Surv. No. Q78/1957/A.

Tinbergen, J. 1958. Unpublished bird report. Brit. Antarct. Surv. No. Q84/1958/A.

Tirrell, P. C., and D. E. Murrish. 1979a. Cutaneous blood flow in the Giant Petrel. Antarct J. U.S. 14(5):169-71.

Tirrell, P. C., and D. E. Murrish. 1979b. Vascular anatomy of the brood patch of the Giant Petrel. Antarct. J. U.S. 14(5):171-72.

Trapp, J. L. 1979. Variations in summer diet of Glaucous-winged Gulls in the western Aleutian Islands: An ecological interpretation. Wilson Bull. 91:412-19.

Trillmich, F. 1978. Feeding territories and breeding success of South Polar Skuas. Auk 95:23-33.

Trivelpiece, S. G., and W. Z. Trivelpiece. 1989. Antarctica's well-bred penguins. Nat. Hist. 12:28-36.

Trivelpiece, W. Z., G. Butler, and N. J. Volkman. 1980. Feeding territories of Brown Skuas *Catharacta lonnbergi*. Auk 97:669-76.

Trivelpiece, W. Z., S. G. Trivelpiece, and N. J. Volkman. 1987. Ecological segregation of Adélie, Gentoo and Chinstrap penguins at King George Island, Antarctica. Ecol. 68:351-61.

Trivelpiece, W., and N. J. Volkman. 1982. Feeding strategies of sympatric South Polar *Catharacta maccormicki* and Brown Skuas *C. lonnbergi*. Ibis 124:50-54.

Valencia, J., J. Janez, H. Nunez, and M. Sallaberry. 1989. Spring and summer diets of pygoscelid penguins at Ardley Island. Unpublished ms.

Voisin, J.-F. 1968. Les petrels ge'ants (*Macronectes halli* et *M. giganteus*). de l'Ile de la Possession. Oiseau 38 (No. special):95-122.

Voisin, J.-F. 1978. Observations sur le comportement des petrels ge'ants de l'Archipel Crozet. Alauda 46(3):209-34.

Volkman, N., D. Presler, and W. Trivelpiece. 1980. Diets of pygoscelid penguins at King George Island, Antarctica. Condor 82:373-78.

Volkman, N. J., W. Z. Trivelpiece, N. P. Bernstein, and P. C. Tirrell. 1982. Macaroni Penguins: Comment on mistaken King George Island breeding record and southerly range extension. Auk 99:386.

Warham, J. 1962. The biology of the Giant Petrel *Macronectes giganteus*. Auk 79(2):139-60.

Watson, G. E. 1975. *Birds of the Antarctic and Sub-Antarctic*. Washington, D.C.: American Geophysical Union.

Watson, G. E., and J. P. Angle. 1966. Unpublished bird log on "East Wind" Cruise, 1966.

Watson, G. E., J. P. Angle, P. C. Harper, M. A. Bridge, R. P. Schlatter, W. L. N. Tickell, J. C. Bond, and M. M. Boyd. 1971. Birds of the Antarctic and Subantarctic Map Folio Series, 14:1-18, plates 1-15.

Williams, J. B., and R. E. Ricklefs. 1984. Egg temperature and embryo metabolism in some high-latitude procellariiform birds. Physiol. Zool. 57:118-27.

Wylie, J. P. 1958. Unpublished biological report, Anvers Island. Brit. Antarct. Surv. No. 11/58.

Young, E. C. 1963. Feeding habits of the South Polar Skua *Catharacta maccormicki*. Ibis 105:301-18.

Young, E. C. 1978. Behavioural ecology of *lonnbergi* skuas in relation to environment at the Chatham Islands, New Zealand. N.Z. J. Zool. 5:401-16.

Zink, R. M. 1978. Birds of the Weddell Sea. Antarct. J. U.S. 13(4):142-44.

Zink, R. M. 1981a. Observations of seabirds during a cruise from Ross Island to Anvers Island, Antarctica. Wilson Bull. 93:1-20.

Zink, R. M. 1981b. Notes on the birds of the Weddell Sea, Antarctica. Le Gerfaut 71:59-74.

Zink, R. M., and J. L. Eldridge. 1980. Why does Wilson's Petrel have yellow on the webs of its feet? Brit. Birds 73:385-87.

GENERAL INDEX

BIRD INDEX

Cormorants: family synopsis, 107; association with penguins, 116; association with sheathbills, 129-30, 132-33
Cygnus melanocoryphus, 40. *See also* Black-necked Swan

Diomedea chrysostoma, 39. *See also* Gray-headed Albatross
Diomedea exulans, 39. *See also* Wandering Albatross
Diomedea melanophris, 39. *See also* Black-browed Albatross
Dominican Gull, 40. *See also* Kelp Gull

Emperor Penguin: Weddell Sea, 4; hypothetical list, 40; sightings near Palmer Archipelago, 47
Eudyptes chrysolophus, 39. *See also* Macaroni Penguin
Eudyptes crestatus, 39. *See also* Rockhopper Penguin

Franklin's Gull, 159
Fregetta gralleria, 105. *See also* White-bellied Storm-Petrel
Fregetta tropica, 40. *See also* Black-bellied Storm-Petrel
Fulmarus glacialoides, 40. *See also* Southern Fulmar

Gentoo Penguin: checklist, 39; status in Palmer Archipelago, 47, 63; boating excursions, 48-49; center of abundance and strategy, 50-51; overall breeding distribution, 62-63; winter records, 66; molt, 66; predation of, 66; breeding biology, 66-68; mentioned, 53, 58, 116, 130, 141, 142
Gray-headed Albatross: checklist, 39; overall breeding distribution, 72; status in Palmer Archipelago, 72
Gray-headed Mollymawk, 39. *See also* Gray-headed Albatross

Greater Sheathbill: checklist, 40; Chinstrap Penguin predation, 61; Gentoo Penguin predation, 66; shag predation, 115; overall breeding distribution, 130; status in Palmer Archipelago, 130; banding, 130; summer records, 130; winter records, 130-31; diet, 131; mortality, 131; breeding biology, 132-33
Gulls: predation, 115; subfamily synopsis, 159

Halobaena caerulea, 40. *See also* Blue Petrel
Hybrid skua: overall breeding distribution, 155; status in Palmer Archipelago, 155; banding, 155-56; summer records, 156; morphology and growth rates of F$_1$ hybrid, 156-57; taxonomy, 157; breeding biology of mixed pairs, 157; mentioned, 148

Imperial Shag, 40. *See also* Antarctic Blue-eyed Shag; King Shag

Jaegers, species synopsis, 135-36. *See also* Skua

Kelp Gull: doctoral dissertation, 23; graduate and postgraduate research, 34; checklist, 40; overall breeding distribution, 159; status in Palmer Archipelago, 160; banding, 160; summer records, 161; winter records, 161-63; distribution of limpets and gulls, 163; foraging behavior, 165; annual diet, 166-67; predatory behavior, 168; mortality, 168-69; breeding biology, 169-71; mentioned, 6, 11, 86, 115, 142, 143
Kerguelen Tern, 173
King Penguin: checklist, 39; status in Palmer Archipelago, 47, 52; overall breeding distribution, 51; mentioned, 39
King Shag, 107

Larus belcheri, 159. *See also* Band-tailed Gull
Larus dominicanus, 6. *See also* Kelp Gull
Larus novaehollandiae, 159. *See also* Silver Gull
Larus pipixcan, 159. *See also* Franklin's Gull
Lesser Sheathbill, 130

Macaroni Penguin: checklist, 39; status in Palmer Archipelago, 47, 68; overall breeding distribution, 68; summer records, 68-69
McCormick's Skua, 3, 40. *See also* South Polar Skua
Macronectes giganteus, 4. *See also* Southern Giant Petrel
Macronectes halli, 39. *See also* Northern Giant Petrel
Magellanic Penguin, 4; checklist, 39; status in Palmer Archipelago, 47, 69; overall breeding distribution, 69

Northern Giant Fulmar, 39. *See also* Northern Giant Petrel
Northern Giant Petrel: checklist, 39; overall breeding distribution, 75; status in Palmer Archipelago, 75-76; diet, 76
Notocarbo bransfieldensis, 40, 108. *See also* Antarctic Blue-eyed Shag

Oceanites oceanicus, 3, 40. *See also* Wilson's Storm-Petrel

Pachytilla desolata, 40. *See also* Antarctic Prion
Pagodroma nivea, 4, 40. *See also* Snow Petrel
Parasitic Jaeger: hypothetical list, 41; records, 136
Penguins: and ice conditions, 11; innocence, 16; site reoccupancy, 16; family synopsis, 47-51; Joubin Islands, 48-49; Gerlache Point, 49; numbering of colonies, 49-50; censusing methods, 50; research in South Shetland Islands, 50; centers of abundance and strategies of pygoscelids, 50-51; association with shags, 116; association with sheathbills, 129, 133; association with South Polar Skuas, 141; association with Brown Skuas, 141, 151, 152; mentioned, xv
Phalacrocorax albiventer, 107. *See also* King Shag
Phalacrocorax atriceps, 6. *See also* Antarctic Blue-eyed Shag